KNOW YOUR DOG

The guide to a beautiful relationship

Immanuel
Birmelin

Hubble & Hattie

1 Recognizing your dog's personality

2 How dogs learn and reason

3

The daily mental workout

Foreword

Each dog has its own sensitive personality

When I was five years old my mother found me one day asleep in the doghouse with three chow chows. She understood the situation completely, and a short time later she gave me my own white chow chow. It was the beginning of a life-long relationship with dogs. Since that time, I have never lived without a dog. As a child, I was drawn to them as if by magic. I understood their feelings and they understood mine. But as I grew older I became more and more interested in how their minds work.

Most books about dogs talk about the history of various breeds and are filled with useful training advice, yet we learn almost nothing about how dogs think and feel. This view has had, and continues to have, consequences: dogs are considered easily trained creatures of instinct. But this way of seeing things falls short of the mark. It doesn't take into consideration that dogs are thinking, feeling beings. Dogs are individuals and have personalities with strengths and weaknesses. I've known this for quite some time, but as a biologist I've never scientifically looked into it until now. I was over 50 before

I started to develop tests that investigate whether dogs are able to reason. Today, there's a boom in the scientific research on the reasoning abilities of dogs, led by the scientists at the Max Planck Institute in Leipzig and the University of Budapest. The findings are fascinating as well as exciting, and they show how far we have underestimated our dogs' abilities.

The success of our film "Which is smarter – a dog or a cat?" ultimately encouraged me to write this book. The aim of this book is to have you get to know your dog from a different perspective, through simple tests that anybody can do at home. You'll learn about, and see for yourself, the mental abilities your dog possesses. You'll understand your dog much better, and your relationship to him will grow even deeper. Enjoy discovering more about "man's best friend".

Immanuel Birmelin

1

Recognizing your dog's personality

A dog is not just a dog. Even dogs belonging to the same breed can't be lumped together under one category. They are each individuals with different strengths and weaknesses, greater or lesser abilities, and distinct emotional lives.

A world of emotions

Emotions are important guides. They determine how we behave in certain situations. Dogs also have emotions. They experience fear, joy, pain, grief and anger, just as we do.

Does a dog have feelings? As a dog owner, you're sure to react to this question with a smile and an unconditional yes. But would you also admit that a fly or a spider has feelings? Most people would say no. When a spider or a fly is annoying, you usually kill it without giving a second thought. Of course, I can't know whether insects and spiders possess feelings, but this example demonstrates how differently we judge living beings. Our sympathies, along with our culture and the way we're brought up, contribute to how close or distant we develop our relations to living creatures.

In some cultures dog meat is eaten with gusto. For myself, being a European and the owner of a Saint Bernard, I cannot begin to understand why these sensitive giants among dogs are raised for their meat in China. The thought of it fills me with rage and anguish. But if I kill a spider or a fly, it hardly affects me. Why do I react so differently, even though I believe

that our crawling fellow creatures also possess feelings? To be sure, the emotional world of a spider is less complex than that of a dog or a human, but this is only half the answer. As a rule, we humans are able to develop sympathy only after we've built a relationship with a living creature. People who raise Saint Bernards in order to consume their flesh don't even begin to do this. For them, these dogs are primarily a source of animal food, in the same way that pigs, cattle and chickens are for us. Whether we attribute feelings to our fellow creatures depends on many factors: our own conscience, our culture and the way we are raised.

Why feelings are important

Imagine not having any emotions, neither pleasant ones such as joy, happiness, and love, or unpleasant ones, such as pain, grief, and hatred. The thought of this makes me shudder. If I didn't have any emotions I wouldn't be able to feel anything. A world without feelings – it's hard to imagine. And yet there are people who are almost totally lacking in feelings. Parts of their brains (particularly the limbic system) have been destroyed through illness. What do these people show us? They're unable to avoid danger or calculate risks, they behave with no consideration for others, and are unable to learn from the consequences of their own behavior. What's surprising is that these people are able to describe their wrong behavior, yet they're not able to deal with it. Thus, they're lacking not in understanding, but in the appropriate manner of using their understanding.

This terrible condition shows us how important feelings are in life. Consciously or unconsciously, they inform the body of its inner world and, by encouraging or discouraging different types of behavior, they help us to make decisions. Feelings are our guides to dealing with the world. It's the opinion of many scientists that all humans of whatever cultural background are born with a "package" of basic emotions. These

A human and a dog: an unbeatable team, even at the computer.

are: fear, joy, happiness, scorn, pain, disgust, curiosity, hope, and disappointment.

Some researchers would add a few more, while others would subtract from this list of basic feelings. Our present emotional life thus consists of a constant mixing of these positive and negative basic feelings.

Personality is shaped by emotions

The comparison with animals comes inevitably to mind. In the wild, living in a complex environment, how can an animal survive if it can't feel fear, assess risks, or feel the pain that will keep it from behaving incorrectly? That's why saying that animals have no feelings doesn't make much sense to me, and even less sense when we consider that their limbic system – the seat of emotions – is well developed. This part of the brain is a kind of transit point where the sensory perceptions of the moment receive their "emotional" expression. This is where the image of Marilyn Monroe is interpreted differently from the image of our mother, where the sound of a bell is different from a siren's wail, and this is where our pain receives its unmistakable emotional quality. This part of the brain is especially well developed in herd and pack animals that actively communicate with each other. Dogs are social animals with a lively culture of communication.

I've lived with dogs nearly every day for the past 50 years, and in this time I've become acquainted with the most diverse types of personality: adventurous, timid, bored, and curious. No two were alike; they were

An affectionate duo: Healthy dogs will always develop protective instincts towards children.

all four-legged individuals with their own strengths and weaknesses. Their personalities were shaped by their emotions. It's not much different from us humans.

The Story of Wisla

Not long ago, I adopted a female St. Bernard from friends of mine. Although they really loved the dog, they had to give it away. All their time was taken up in caring for their mentally and physically disabled son. They had secretly hoped that the dog might be able to help the boy, but unfortunately this was not the case. The boys disabilities were simply too great. Wisla – the name of the year-and-a-half old female – was indeed very sweet with the child, but she didn't make things easier. On the contrary, being a young dog, she wanted to play and discover the world, and my friends weren't able to give her the opportunity to do so. There was simply no time for playing and going for walks

◆ **Love at second sight:** My first encounter with Wisla was anything but cheerful and friendly. There wasn't a trace of love at first sight. We met in the apartment of a mutual friend –for Wisla, foreign territory. The dog and her owners had just completed a

TIP

Getting to know each other

If you're bringing home an older dog, take its past into consideration. The dog has to make adjustments. It has to become familiar with new surroundings and strange people. Don't overwhelm the animal with your feelings and expectations. Give it time. If it misbehaves, don't react immediately with scolding and punishment. Wait until the dog has established a bond with you before you start using rewards to change its behavior.

long car trip from Denmark to southern Germany. Perhaps they were all simply tired and worn out, but I don't believe this explains Wisla's behavior. She bared her teeth and growled at me, and the fur on the back of her neck stood on end. I started talking to her and tried to win her over with a treat. No chance! What could I do? I could actually see what was going through my friends' minds. They were thinking I wasn't the right person for their Wisla. Our "chemistry" didn't seem to click. My experience with wild animals had taught me that patience was called for. The thing to do was to wait and calmly observe the animal. I continued talking to my friends without letting Wisla out of my sight. These were our first sobering hours together.

◆ **Breaking the ice:** On the next day we all went out together for a walk. Wisla now knew me, and I waited until she took the initiative to make contact with me. I didn't want to push her in any way. It was playing with a stick that broke the ice. She was in a playful mood, and that helped me in approaching her. I deliberately didn't tempt her with food because I wanted to see her in action. We were completely relaxed as we ran around in the grass. She forgot herself, the way children do when they're playing. Our first walk cracked open the window to her emotional world. I saw that when encountering other dogs she was quiet, but at the same time cautious or even afraid, while with humans she reacted joyfully or indifferently. She adored children. I sensed her strong fighting instinct from the way she struggled with me over

Not every fellow dog is welcome. Just like people, dogs also have their likes and dislikes.

the stick. She didn't let go until it belonged to her. I could tell how attached she was to her owners, since she would always go back to them wagging her tail. She showed me many facets of her personality that appealed to me. Thank God! After all, for things to work out I also had to have the right feeling about her. Otherwise, there would be no chance. She would have detected the slightest dislike. A first impression may often be correct, but in human-animal relations one needs greater certainty.

◆ **Feelings need time to develop:** During the following days and weeks I keenly observed the new member of our family. Wisla was obeying either badly, or not at all. My family was putting pressure on me to teach her obedience. What to do? There's a commonly held belief that a dog has to be shown right away who the leader of the pack is, but I don't agree this. And I also don't believe, as many people who hold the first belief do, that the dog will later get away with anything it wants. Some people just don't stop to consider that feelings need time to develop. The time to begin with simple obedience training doesn't come until the dog has developed feelings of trust, security and affection (→ pgs. 44-45).

Now the dog and I understand each other better. I can now see if she's afraid, shy or courageous, and I can deal accordingly with her basic feelings. Trying to raise a timid dog by being domineering is the wrong approach. It will only take away even more from its self-confidence. Instead, it needs to be challenged, in the best sense of the word.

Feelings are different in each individual

We share with dogs many of the same emotions. And jealousy, grief, joy and love differ from dog to dog just as they do from person to person. Probably no two living beings feel exactly alike, only approximately so. You would think this is obvious, yet we set higher standards when it comes to our pets. Since we can't directly measure their feelings, we either play them down or exaggerate them. Neither attitude is conducive to a better understanding of our pets. It's important for me to know if my dog is jealous, for example, or if he can easily handle this feeling. A brief story demonstrates this point.

◆ **Robby and Teddy:** Robby, a sweet, loveable golden retriever, lived for several years with Teddy, a long-haired sheepdog, until Teddy died. The two dogs were very different in their emotional makeup. In contrast to Robby, Teddy hardly knew the meaning of jealousy. He knew he had a place in my heart. Robby, on the other hand, would react with loud barking to any imagined slight. When greeting me he always pushed ahead of Teddy. Luckily, this didn't bother Teddy in the least. He was the image of

calmness and the boss of the dog community. After getting his share of affection, Teddy would withdraw. Not so with Robby. He couldn't get enough of it, and that was fortunate for us and the dogs because the difference in their characters allowed them to live together without quarreling.

◆ **Robby and Wisla:** Co-existence for Robby and Wisla, the Saint Bernard female, took a totally different form. At first sight, things should have been more peaceful since, as a rule, a male and a female dog get along well together. This was, however, not the case here. Each dog craved human affection and was jealous of the other. One day, upon my

TIP

Dogs are individuals

A dog's distinctive personality is more important than the characteristics of its breed. A timid dog, for example, requires a great deal of encouragement and attention. An adventurous dog, on the other hand, has to be reined in every now and then. In training a timid dog, you first have to take away its fear. The daredevil, who will do anything to get its own way, needs rules that should be taught lovingly, but also firmly and consistently.

return from a long trip, both dogs were happy to see me again. Robby was barking as usual, wagging his tail and pushing past Wisla, the way he always does. Wisla's reaction was quick. She snapped at Robby and attacked him. I automatically stepped in and yelled at her. She became frightened and left Robby alone. So far, so good. But it's important not to let animals stay with bad feelings, since this can later develop into hostility. So how did I handle it? After the quarrel I called the two dogs to me on either side, petted one on the left and one on the right, talked to them and hugged them.

My aim in doing so was to give Robby a feeling of safety and belonging, and to show affection. And Wisla? She should sense that we loved her and would never reject her, but that she should not attack Robby. It's sometimes difficult to look into an animal's mind or to understand our fellow creatures better, but it's worth it

◆ **What's important for you and your dog:** Dogs, like all living creatures, exhibit behavior typical to their species. Many of these behavioral characteristics are genetically determined. Dogs, as a rule, do not like living alone. Their genes tell them to live in packs, where they have to, and want to, follow certain rules. A human replaces the pack. Their genes also make them wag their tails when they're happy, bark when they're excited, and so on. These inherited behaviors show us the boundaries of a species. To overstep these boundaries is cruel to animals. Forbidding a dog to use its nose would be such a case. But dogs are also individuals. There is no "typical" dog.

1

Happy and alert

Dogs also have strong feelings which you can see in them if you know them well enough. Through her posture and facial expressions, Aisha is showing me that she's happy. Maybe her owner is beckoning to her with a treat.

Dogs are all different, even though they share common attributes. The key to a dog's personality is knowing the dog's feelings. Think of this concept like a painting: The picture frame establishes the boundaries of the work of art, but the picture itself is created through the art of painting.

2

3

Tense and concentrated

A dog's exceptionally keen hearing hardly misses a sound. Aisha has heard something rustling in the grass. Could it be a tasty mouse? She waits tensely at the spot where the noise is coming from.

Totally enraged

When her mistress is around, Aisha will also let herself get furious with a dog that she doesn't like. Growling and barking, she shows him her fangs. Her tail is pointing down. The message translates as, "Get the hell out of here!".

How to recognize your dog's feelings

At first, it may seem very difficult to recognize a dog's feelings, since its emotions are located in a world hidden from us. What connection is there between the inner and outer worlds?

Fortunately, there are many types of behavior that reveal a dog's feelings. As the famous behavioral scientist and Nobel prize winner Konrad Lorenz has described, a dog's grief is easy to recognize. The dog becomes quieter, its eyes have a different expression, and the positioning of the ears changes. A dog that cowers and pulls its tail in is afraid. If it's baring its teeth and the fur on the back of its neck is standing on end, it's being threatening. These types of behavior are familiar to everybody.

A dog's facial expressions, gestures, and vocalizations communicate its feelings to us. This becomes more difficult when a signal cannot be assigned with certainty to an emotion. Baring teeth does not necessarily signal an attack; it can also express insecurity. The photos above show different types of behavior such as joy, concentration and anger.

Can you influence your dog's feelings?

Aggressive male dogs are usually neutered. The aim of this is to make the dog peaceful, but this can also often be achieved by correct timing.

◆ **The effects of neutering:** Neutering, or castration, involves the removal of the testicals, causing the body to cease production of the male hormone testosterone. Large amounts of testosterone in the blood increases aggressiveness. Even male lions will turn into kitty cats after being castrated. The temptation to neuter aggressive, dominant male dogs is understandably great. But this is only one side of the story. The animal's behavior changes fundamentally, and its personality changes significantly. Many positive characteristics are gone forever. Thus, for example, its concentration and fighting instinct are lessened. Unfortunately, such dogs are also prone to obesity. That's why my advice is to resort to neutering only when all your efforts at training, understanding and engaging your dog have not helped.

◆ **Unhealthy emotions:** Pathological anxiety, fear and compulsive neuroses are caused by abnormal brain processes. The balance of certain chemical substances arising from under or overproduction, is often disturbed. In serious cases of mental illness this can lead to a total collapse of the brain's chemical framework. In this case, only medication can help. Its effectiveness was shown to me by Karen Overall, a noted American veterinarian. For years she has been a specialist in psychological illness in animals, and is an authority in this field. She introduced me to Cody, a male Dalmatian, which one day suddenly began to run in circles, biting its tail. As if that weren't enough, when his owners tried to stop him from doing this, he became threatening. Karen Overall administered medication developed for humans with compulsive disorders. This was also effective on Cody. Since then, Cody has been taking his daily tablet and his life is almost as normal as before. Cody hasn't been cured, but with the tablet he's able to live normally. Cody is not an isolated case. More and more dogs suffer from psychological problems. An estimated 15 percent of all dogs suffer from a fear of separation. When their owners are away

INFO

Compulsive behavior

If your dog, let's say, repeatedly runs around in circles trying to bite its tail, this could be the beginning of a neurosis. In such cases, don't just watch and be amused. You should interrupt this behavior by distracting the dog, perhaps through a familiar sound such as rustling paper, or by giving it a treat and speaking to it soothingly. This way you can prevent these incidents from becoming ritualized behavior.

This dog's owner needs some special comforting today. This much we know: No animal understands a human's emotions as much as a dog.

they start panicking, howling, yapping incessantly, and destroying or dirtying their homes. In very serious cases a veterinarian will administer medication that will effect the animal's mind. Of course, the best course of action is to get a second dog, since being alone is difficult for dogs, which are pack animals.

Fortunately, not many dogs require medical intervention and medication. The majority of pets are content to have humans as their pack mates.

◆ **Dogs and people understand each other:** A dog and a human can sense each other's feelings. They live in a type of emotional symbiosis. There is no animal on Earth that understands a human's emotions as well as a dog does. This is not a coincidence, but rather the result of thousands of years of living together and of breeding. Humans have created the dog in their own image. It is therefore no wonder that through skillful guidance humans are able to influence the emotions of their four-legged friends. Excessive jealousy and anxiety can be alleviated through plenty of patience and knowledge.

Emotions can be guided

Jealousy, fear and grief are strong emotions. Here are three practical examples of how you can deal with your dog's emotions.

■ Jealousy

Ever since the baby came, our dog has been behaving strangely. He constantly wants to be petted and his shoving makes him a nuisance. What can we do?

A baby unconsciously becomes the center of everybody's love and attention, and that's just fine and wonderful. However, your pet won't understand this. He doesn't know the baby. To him it's something foreign, and he first has to get used to the new person and the situation. How can you make the adjustment easier for him? By holding the baby in your arms in the presence of your dog and showing it to him. While doing so, don't forget to talk to him and the baby. Gently stroke them both. It's good for them, and the dog will learn that he's not being ignored, but that he has to share his love. This process takes a bit of time and patience, but when you succeed in communicating warm feelings to your dog despite the baby, you'll have found a friend and protector for your child. You don't have to go as far as John Aspinal did when he let his grandchildren play with gorillas. I myself have seen how gently a female gorilla took his grandchild in her arms and romped with it. I asked John Aspinal and the parents if they were afraid for the child. The quick reply was, "Gorillas live in groups and are extremely gentle with their young. They can also see how fragile a human child is." Douglas Hamilton, who does elephant research, walked with his baby among wild elephants, and the elephant cows cheerfully sniffed the child. These examples show that highly social animals are able to deal with the offspring of another species when you give them the opportunity to get to know them. That's why my special advice is to let your dog smell your baby from head to foot. You can forget about hygiene for a few moments.

■ Insecurity/fear

A short time ago Lucky, our Dalmatian, was bitten by a much smaller dog. Since then, Lucky has been afraid of other dogs. If we're taking a walk and we meet an unknown dog, Lucky will freeze in his tracks and pull his tail in. How can I take away his fear?

Fear can be conquered only through having a lot of positive experiences. That's why you should have Lucky meet a lot of peaceful dogs. The best thing is to find playmates for him. These encounters will teach him that many of his fellow dogs can be peace-loving friends. And that's not all: He'll also become a better judge of other dogs' expressions and gestures. This knowledge will make him more secure in his dealings with other dogs and will gradually take away his fear. In my experience, an especially timid dog will never totally lose his fear, but it can still be lessened.

What's up? Aisha is alert and paying attention to something interesting near her.

■ Grief

Not long ago Maxi's owner died and we adopted him. But nothing can cheer him upi. He just lies apathetically in his basket, hardly eats, and drags himself over when we call him to come. How can we help him get over his grief?

Nothing doing. Aisha is bored. She's resting her head on her paws and waiting to see what happens.

It's very difficult to help people who are grieving. You can give comfort to a grieving person by sharing your feelings and sympathy, but however cruel it may sound, you can't actually help anybody. People have to work out their own grief. It's even more difficult to help an animal get through its grief, since it can't communicate through speech. I think the best thing to help Maxi is another dog whom he could, and would have to, get used to. Perhaps this will help him to forget his owner. Other dogs are good catalysts for a dog's mind. Every walk you take where Maxi meets other dogs and can move freely will let him forget a bit of his grief.

Learning, reasoning and feeling

These three words will be repeated throughout this book. Learning, thinking and feeling form one entity and are linked together in different ways.

Learning

With learning it's quite apparent that thinking and feeling also play important roles. It's easy to teach an animal tricks by rewarding it with food. This is nothing other than satisfying an elementary need, namely, hunger. To satisfy feelings of hunger, humans and animals will do almost anything.

It's a basic rule that the driving force, the reason why humans and animals learn new things, is the expectation of a reward. A reward can be more than simply food.

A highly intellectual pleasure, such as understanding a difficult mathematical process or experiencing a piano recital, is also coupled with a range of feelings such as excitement or delight. It's nearly impossible for a higher animal to learn without nuances of feeling.

Reasoning

And what's the situation with reasoning? We all know how difficult it is to solve a riddle when we're afraid. The feeling is usually too strong, and we can't even begin to concentrate. It's pretty much the same with animals. Before they start solving a problem, they have to be free of fear. This has been shown in many scientific studies. Too much stress leads to skittishness and the restriction of thought processes and behavior.

Feeling

Positive feelings are helpful not only for people but also for animals when they're solving a problem. A requirement for our intelligence tests with dogs was that the animals had to be free of stress and anxiety (→ daily mental workout, pg. 66).

A cool reception. The older dog isn't pleased at all with the puppy's friendly overtures.

20

A dog's senses

Dogs have different sensory perception than we do. Their sense of smell, for instance, is much better than ours. With their "supernoses" they can notice things in their environment that remain hidden to us.

Dogs perceive the world differently than we humans do. It's incredibly difficult for us to understand or grasp this fact because we ourselves are prisoners of our sensory organs.

We will, however, have to plunge into a different world if we wish to understand our pets. In not knowing how our dogs perceive the world, we could misinterpret their behavior and demand too much from them. A small example can illustrate this point.

Try to train a dog to fetch an object out of a red box, from a row of colored boxes. It's senseless, since dogs can't see red. They can't tell the difference between green, yellow, orange and red. They don't see the colors in a traffic light. They recognize that the red light means stop and the green light means go only by the brightness and the position of the lights. People who suffer from red-green color blindness perhaps see the world in a similar way.

When shown a card with red and green spots they are unable to recognize the red or the green, yet they see the world in colors. Dogs may not be visual animals the way we are, but instead they are masters of smell. No other domesticated animal possesses such a fine sense of smell.

How can you discover your dog's sensory world?

At first it may seem easy to see the world through the eyes of a dog, but the difficulty lies in the details. To determine exactly how well a dog can smell, see and hear requires a well-planned series of experiments. Let's just stick to one method, that of learning through differentiation. If you want to test which colors a dog can recognize, follow the test described on page 27. Similar tests explore the other senses (→ pgs. 29 and 31).

Why it's important to know about your dog's senses

Knowing something about how your dog senses the world will make living together easier. Very often we're annoyed by barking and yelping. We're puzzled and we get angry because we don't know why the "stupid animal" is making so much noise. In reality, the dog is hearing or smelling something that it perceives as threatening, but that we haven't noticed. We don't understand its good intention of protecting house and home because our senses can't perceive the possible danger. Some of our pets' unaccountable reactions of fear and alarm can thus be explained simply. They're hearing high-pitched sounds that our ears can't hear. In the course of millions of years, each species has developed sensory organs that allow it the best chances for survival. A good example of this was the catastrophic tsunami in east Asia. The tidal wave killed far fewer large mammals than humans. Elephants, for instance, were able to sense the coming wave and fled to higher ground.

Dogs are masters of smell. They "see" the world through their noses.

Through the eyes of a dog

How do dogs see? Light enters through the dog's eyes and is absorbed by the sensory cells – the rods and cones – located in the retina. These sensory cells transmit electrical signals to the brain, where an image appears.

Then things get interesting, since the distribution of sensory cells in the retina varies. This means that vision is not equally good in every area. An animal has the sharpest vision in the areas where the sensory cells are arranged most densely. It's much like a digital camera – the more pixels, the sharper the image.

A dog's retina, in contrast to a human retina, has two favored areas where the sensory cells are densely packed. One is called the area centralis. Here, the sensory cells are arranged in circles. This is in contrast to the horizontal bands, where the sensory cells are placed tightly next to each other, like paving stones on a sidewalk. Until recently, that was the extent of our knowledge. In the past few years, however, Australian researchers have gone a bit further down the path of revealing this mystery.

Alison Harman, a neurologist, made a surprising discovery. After examining the retinas of different dog breeds she found out that many short-muzzled breeds don't possess any horizontal bands with rods. Long-muzzled breeds, on the other hand, had the horizontal bands. This was an enormous surprise, and it explained why some dogs have a wider field of vision than other dogs

The fascinating blue eyes of a Siberian husky. Dogs see the world differently than humans do.

Short-muzzled dogs "see" better

What does this mean, practically speaking? Dogs with long muzzles and horizontal bands can see peripheral objects more easily. That's why they're also better hunters, since they can recognize prey on the horizon. It's also no coincidence that hunting dogs have a longer muzzle.

However, dogs without these bands (i.e. with only an area centralis) are able to recognize better the nuances in the faces of their owners. That's because the area centralis of a short-muzzled dog contains a great deal more sensory cells than that of a long-muzzled dog. Perhaps that's the reason why we find bulldogs, boxers and other short-muzzled dogs so attractive. They just have to look us in the eyes and our hearts melt.

This also incidentally answers an old question, namely: Why do some dogs like watching TV, while others don't? It's a lot more fun for our cuddly bulldogs because they can see television images much better.

Sharp vision

How sharply we can see a picture or an object depends, among other factors, on how dense our sensory cells are arranged. The density is much greater in humans than in dogs, so we see objects more sharply. What does that mean? A human, for example, can recognize two small stones of the same size, spaced apart at a quarter of an inch, as two separate stones when viewed from a distance of twelve feet. Dogs, however, are unable to differentiate so precisely. For them, there's only one small stone lying there. They can see two stones only when they approach to a distance of six feet. From this, it follows that a dog's vision is 50 percent less sharp than a human's. Thus, you can't expect a dog to recognize you when you're standing quietly at the edge of a field at a distance. It can't see you. It won't recognize you and come running until you start waving your arms. Dogs find it much easier to see objects that are in motion. Remember this when you want your dog to come to you from a greater distance (→ the vision test, pg. 27).

Stereoscopic vision

A dog's field of vision comprises 150 degrees and is noticeably larger than that of a human. This means that a dog will see an approaching object sooner than we will. This has its price, however: It's stereoscopic (three-dimensional) vision is worse. This field of vision is only 85 degrees, in comparison with 120 degrees for a human. Stereoscopic vision is naturally of great importance to human beings, since it helps them estimate distances. This ability has made humans the toolmakers among mammals. You can test for yourself how important this skill is. As an example, simply try to thread a needle with only one eye open!

TIP

Dogs see the world differently than we do

Dogs, which are by nature cautious, will often approach an unknown object, such as a crow or a felled tree trunk, with aggressive barking. Such behavior is inexplicable for us, since we don't notice any threat. When something like this happens, just lead the dog to the object and let it sniff at it thoroughly. This way, the dog will realize the object poses no threat or danger.

Twilight vision

In this respect a dog's vision is superior to our own. Its lens and cornea can take in more light in dim surroundings, and the back of the eye is covered with a layer of cells (the tapetum) that reflects remaining light like a mirror. The ancestors of today's dogs were thus able to hunt well in twilight. You will notice that dogs like chasing after animals especially at dusk. They can see their prey, but their prey can't see them

Color vision

We're certain that colors don't play an important role in a dog's world, yet it's not quite that dogs are color blind, as was previously believed. They don't have sensory cells that register the color red (→ pg. 22). This brief excursion into the visual world of dogs shows how differently from us they view the world. And again and again they surprise us.

How does a dog know that it's a dog?

This is exactly the question we researched in a two-year study. This may sound absurd and, after all, what does it have to do with a dog's senses? Yet in my opinion the question is not so far-fetched. Ultimately, there isn't an animal species that takes so many different phys-

ical forms as the dog. A Saint Bernard, for example, looks totally different from a miniature pinscher. But in our research we came to no definite conclusion, yet there is still reason to believe that dogs recognize each other through their physical appearance.

It was almost as a by-product of this study that we made a surprising discovery. My own German Shepherd didn't recognize me anymore when I put a bucket over my head and ran around on all fours in front of him. His reaction was not harmless. He tried to bite me. This came to me as a total surprise, since everyone in my team thought he would smell my familiar scent. But when I raised myself onto my knees in front of him –my head still in the bucket – he knew right away who I was.

◆ **Altered appearance:** We tested several dogs individually with their owners and a familiar person. The three were left in the testing room busying themselves. After about ten minutes, the familiar person led the dog out of the room while the owner put a bucket over his head, and was asked to get down on all fours. This process took about two to three minutes. Then the dog was brought in, and its reaction was written down and recorded on video. Each dog's reactions was different. Some of them tried to attack, while others were frightened, but each one recognized its owner with the

The vision test

Test your dog to find out what colors he perceives and from what distance he can see you. You can perform both these tests.

■ The color test

Preparation: Put one green and one blue dog bowl in front of your dog, with food only in the green bowl, not in the blue

Performance: After feeding five to ten times, the dog has learned that he'll find food only in the green bowl. You might now object that the dog can smell where the food is. As a control, put the food into the blue bowl. If the dog goes for the green bowl, you can exclude the case for smell. Now switch the position of the two bowls to make sure that the dog isn't being led by their placement. The situation is now more difficult, since each color has its own shade of gray. What this means will become clear to you if you lower the color level on your TV screen. It could thus be the case that what the dog sees aren't colors, but shades of gray, as was long believed. However, further, more refined, tests have shown that dogs do see colors – perhaps not the bright colors that we see, but colors nonetheless.

■ The distance test

Today we have exact figures for how well a dog can see from great distances. A German Shepherd cannot recognize a person who is standing motionless at a distance of 500 yards, but will have no problem in recognizing someone who is moving at a distance of 800 yards. I tried testing this result with my own dog.

Preparation: I performed the following test with two good friends who my dog knows very well: While going for a walk my two friends slipped away unnoticed, and after five minutes reappeared, standing motionless, at a distance of 450 and 800 yards.

Performance: My dog had absolutely no reaction to either one, even though he was looking in their direction. It wasn't until the more distant person started waving that the dog wagged his tail and whimpered. His behavior also spoke volumes. He happily ran towards my friend and didn't even glance at the person standing nearer. What would your dog do?

bucket on his head when he kneeled in front of them instead of walking around on all fours. Perhaps this experiment explains how a dog will suddenly attack a child. While playing, a child happens to put something over its head. The dog doesn't recognize the child anymore and attacks. We can only speculate on what the dog sees or perceives in this concrete situation.

◆ **Careful!** You should never try the experiment described above on your dog. There's no way of knowing how it will react.

Through the nose of a dog

Humanity has long made use of a dog's nose, in earlier times as a tool for tracking, and today as a sensitive detector of smells.

Here, all the sniffing and digging has paid off. This dog heard a mouse scurrying.

Dogs help humans in detecting illegal drugs, in medicine in the early detection of bladder cancer, in the search for missing persons, and in fighting crime. A few figures will give you an idea of how superior dogs are to us in the realm of smell.

The dog: a "supernose"

A human has about five million olfactory (smelling) cells, while a dog has 200 million. The olfactory center in its brain is seven to ten times larger than that of a human. One third of a dog's brain processes signals from its nose, compared to only one twentieth of the brain in a human. To give an example, a German shepherd needs 500,000 molecules of acetic acid per milliliter of air in order to smell the acid, where a human needs 50 trillion molecules.

There are differences in smelling

Smelling ability differs from breed to breed. German shepherds are among the champions. Among them there are specialists that can even smell molecules (substances) that have penetrated a shoe sole. Almost nothing escapes a dog's nose. Admittedly, not all substances can be smelled equally well, but it is certain that the sense of smell is used by dogs for communication. The details of what it can communicate by smell are largely hidden from us. Perhaps it's a type of code language. Dogs have developed a special organ – the Jacobson's organ – for sexually attractive scents. It's located on the floor of the nose. The information saying that a female smells good is sent directly to the

The "supernose" test

Dogs enjoy doing nothing so much as exploring their surroundings with their noses. Find out how good your dog's sense of smell is.

■ The salami test

1. Preparation: Give your dog plenty of time to sniff at a piece of salami. Then take him out of the room. Drag the salami across the floor to create a scent trail and then hide it. Bring the dog back into the room and encourage him to look for the salami

1. Performance: The dog doesn't hesitate in tracking down the scent.

2. Preparation: Things now get somewhat more difficult for the subject. The experiment is the same, with only one difference. Have the dog sniff at the salami again. After he's left the room, drag a piece of cheese over the salami trail. Have the cheese and salami trails branch apart after a few yards.

2. Performance: The dog follows the salami trail. In this experiment, we humans would have had no chance, since we would only have detected a "hash" of smells.

3. Preparation: The experiment starts off as usual. But this time, instead of on the floor, rub the salami onto the sole of your shoe and start walking.

3. Performance: At first, the dog eagerly sniffs the trail of your steps, but then doesn't follow it. Why not? Dogs have the ability to recognize the concentration of a substance, and will always follow the higher concentration. Since the concentration of the scent naturally diminishes with each step, the dog remains at the starting point.

■ The human scent test

Preparation: Take a fork (or a spoon) and rub it well against your underarm. Have a friend do the same with another fork. It's important that both forks look exactly alike. Hide the forks in separate places so that the dog can't see them.

Performance: Have the dog sniff your friend's underarm or a piece of his clothing for a minute or two, then immediately have the dog look for your friend's fork. A skillful dog can easily differentiate your scent from your friend's scent, and he'll find the right fork. Just what is your dog smelling? He's smelling skin cells that have been left on the fork.

Almost nothing escapes a dog's hearing. It senses even the softest sounds.

brain's limbic system, the place where emotions arise.

Through the ears of a dog

A dog's ears don't make it easy for a small mammal such as a mouse to escape. There's not much that escapes a dog's hearing organs, which are like moveable radar dishes searching space for signals. Its hearing is much keener than a human's. Dogs can locate two sources of sound from a distance more easily than we can. For us, the two sources of sounds would sound like a single one. A dog is able to pinpoint a sound ten times more accurately than a human. This makes tracking prey easier. Dogs are also helped by their ability to pick up sounds in the ultrasound range. They hear frequencies ranging from 20 to 60,000 hertz. We humans hear in the range of 15 to 20,000 hertz. That means they can hear high-pitched sounds that are hidden from us. That's the reason high-frequency whistles are often used to train dogs. Obedience (or training) schools exploit our pets' good hearing when they use clickers. The popularity of clicker training is due to the fact that it makes shouting commands unnecessary.

◆ **Hearing problems:** Dogs can also become hard of hearing as they grow old. You can easily find out if that's the case with your pet. The following exercise is not a medical test, but it will give you an idea of the current situation. Have your dog sit in a room while you're standing behind him about six feet away. He shouldn't be looking at you, and you should behave quietly. Then start making noises using different objects. If he can hear you, you'll notice him pricking up his ears and turning his head. If he doesn't do this, he has a problem.

A dog's sense of taste

Dogs are not gourmets, which is evidenced by the number of papillae (sensory cells) on their tongues. They possess 1,700 of them, while humans have 9,000. But they don't ignore taste, either. Some dogs prefer cooked meat to raw, and they enjoy a sweet dessert as much as we do. They also have papillae for sweetness, but don't feed them sweets, since they're harmful to your pet. Apparently, dogs can differentiate among salty, sour, sweet and bitter tastes. Just how well and how intensely they do so remains a riddle, however.

The hearing test

Dogs can hear sounds at ultra-high frequencies that we can't perceive. They also hear certain frequencies better than we do. Test this yourself.

Our ears react most sensitively at 2,000 hertz. This is where we best hear a soft sound. In dogs, the greatest sensitivity is at 8,000 hertz. At this area of frequency we already have to speak much louder for others to hear us. Our sensitivity has diminished noticeably. Just what this means for dogs and humans has been described impressively by Stanley Coren in his latest book. When pronouncing words such as "show", "shine" or "shut" and drawing out the "sh" sound, most people do this at a frequency of 2,000 hertz. But when one pronounces the sibilant "s" sound, which sounds like the buzz of a hornet, this occurs at about 8,000 hertz. To our ears, this sound is softer, but it's quite the opposite with dogs. They register it as being louder. Try this with your dog and see.

■ The sound test
You can determine for yourself how well your dog reacts to certain sound frequencies. Teach him to come and get a treat from you with a certain sound. How can you do this? It's easy.

Preparation: First, give your dog the "down" command, then walk ten to twelve feet away and stand facing him.
Performance: Now use a recorder, flute, or similar instrument to play a certain note. After you've done this, call him to you and reward him. After this is repeated a few times, the dog has understood the exercise and knows that he's supposed to come to you at the sound of the note.
Increasing the level of difficulty: It's also not hard for the dog to recognize the difference between two or more notes. Along with the note he's learned, play another one, and this time pet the dog afterwards. You'll see how well dogs can differentiate between sounds. Incidentally, here's a funny story about my dog Teddy. When I did this experiment with him everything went fine, except for one high-pitched note. When I played it to him, he started howling like a wolf.

2

How dogs learn and reason

Nobody doubts the fact that dogs are able to learn. But for a long time the idea that dogs are able to reason was disputed. Now that it's been proven, we see our beloved pets in quite a different light.

The way your dog learns and reasons

"Learning for life" is an old saying. This applies to humans and dogs alike. Learning continuously means keeping fit mentally and being able to handle problems successfully.

Why do we learn? The answer sounds simple, yet baffling: for the joy and pleasure of it, or perhaps to get a better job. Becoming a good skier or pianist first requires a great deal of practice. Only later, with increasing progress, does joy enter into it. Learning can thus be an investment in future "happiness" or in a secure professional career. The better your education – i.e. the more you've learned – the better your chances in the job market. Humans are surely the pinnacle of creation, at least as far as learning goes.

And why do animals learn? For reasons similar to those of humans. For them too, satisfying a positive feeling or avoiding a negative one is the driving force. A human's career corresponds to an animal's struggle for survival in the wild. Learning enables animals to adjust to changes in their environment. This is how they become fit to meet the chal-

lenges of everyday life. Learning is thus a universal principle in nature. Just what, and how much, can be learned varies from species to species. Dogs, for example, learn quite easily to understand signals from humans. This is an important requirement for training an animal. Dogs are champions in understanding these signals, which is why it's no wonder they help us in almost all walks of life. They are companions for the blind, they sniff out narcotics, they hunt with us, they comfort us, and they do much more besides. These gifts have made dogs our friends and helpers. But this hasn't always been the case …

Dogs enjoy learning

The story begins 50,000 to 100,000 years ago. Exactly when is still a matter for dispute among scientists. Whatever the case, what's certain is that man at one point began breeding dogs according to his own wishes. The result was an animal with an extraordinary gift for learning.

34

But being gifted alone is not enough. As the saying goes, "No pain, no gain". Of course, dogs aren't hard working in the same sense as humans are, but they're often tireless in the way they learn new things. A dog will continue trying to learn a task that a cat would have long given up. Perhaps that's the reason cats are so difficult to train. With a dog, it's always possible to repeatedly practice a difficult learning task. They have stamina. Without this attribute it would be impossible, let's say, to train a seeing-eye dog.

A seeing-eye dog needs the abilities of comprehension, stamina and persistence.

Months of intensive training are required. A crash course is useless. This is a difficult period for both dog and trainer. Each depends on the other. The trainer needs to be patient and sensitive with his animal if he wants to be successful. Success also means that the dog is enjoying his work. And now we come to something surprising, and even unfamiliar to many people: dogs enjoy learning – their brains practically demand it.

◆ **A brain can be trained:** It's only been for the last few years that we've determined that the brain can be "redesigned" to a certain extent through training and experi-

Dogs want to be challenged. A little slalom run keeps this dog fit.

ence. It's similar to a construction site where scaffolding is constantly being built, new lines are being laid, and old ones are torn out. Animals that learn continuously and live in a stimulating environment have nerve cells that are longer and more branched out than is normally the case. The discovery that training leads to nerve growth caused a scientific sensation. Not only does learning and training "lubricate" the switching circuits in the brain – it even creates new switching circuits and nerve connections by retrieving genetic programs that renew and extend the nerves. The brain is an organ that, within given limits, renews itself through use – that is, through the daily confrontation with its surroundings.

In a stimulating environment, dogs are more alert, more curious, and also mentally more agile. Learning means building new nerve connections, and many learning processes appear intelligent to us at first sight. A dog is considered smart if it brings its master the newspaper. But in reality it hasn't grasped the meaning of its actions and doesn't know what it's doing. What it's learning is simply, "If I bring the paper, I'll get a treat."

Learning can thus function without reasoning. When a dog or a human has learned a task, it's often then performed in a robot-like fashion. Anybody can remember how difficult it was to learn a language or to ski well. When you're speaking fluently or whizzing down a slope, you're not thinking anymore about grammar or how to shift your weight on the skis. Everything has fallen into place, and thinking about it

Time for a walk. Lucky knows exactly what's happening, so off he goes to fetch his leash.

would even be an obstacle. If you started doing so, you'd begin to stutter when speaking a foreign language, or fall down while skiing.

◆ **What's important for you and your dog:** The point of learning is to adjust quickly to specific changes in the environment. This gives the dog the possibility to tackle new challenges. Dogs like learning, and nothing is as bad for them as boredom. Learning can take place without reasoning.

Are dogs able to reason?

Until now, reasoning had been considered a privilege of humans. Only in the past few years have opinions turned in favor of "canine intelligence". A dog's powers of reasoning have now become a subject for scientific research. The surprising results haven't reached dog owners yet, but that's going to change. From weak-willed, obedient servants, dogs will become thinking partners. We'll see our friends in a different light, not as a version of humans, but as dogs with individual qualities specific to their species. For eight years I've been doing research, together with a small team, into the reasoning abilities of dogs, and for me there's no doubt that dogs are able to think logically.

INFO

Learning and reasoning

A dog's brain wants to learn and figure things out. A lack of stimulus will cause a dog, especially a young dog, to "transfer" his boredom into unwanted types of behavior. For example, it will start chewing, or even destroying, furniture. That's why you should stimulate your dog into thinking and learning, and make sure it experiences new sensations. Also let it play often with other dogs.

What is reasoning?

For me, reasoning means playing out a situation in your mind. We act out different scenarios in our minds in order to judge how they'll turn out. This idea is illustrated by a chess player: In his head he plans his and his opponent's moves. The further ahead he can think, the better he plays and the closer he gets to winning.

Reasoning and learning support each other to a certain extent. This means that certain problems can be solved more easily with the knowledge acquired through learning. We've learned so many countless things that we often confuse reasoning with learning. This becomes suddenly apparent when a situation that we've learned to deal with has changed.

At an exhibition in Basel I saw just how desperate someone can become in such a situation. The task at hand was to use a wrench in order to insert an oversized screw into a specially sized hole. The problem was that the screw wouldn't go in when it was turned in the usual fashion. The solution was in turning it counterclockwise. More than a few people didn't pass the test.

Why do animals reason?

For animals, the world is full of problems: escaping from enemies, obtaining food, raising young, and much more. Survival means overcoming these problems, and nature demands a high price for this. Thinking doesn't come cheap, as far as energy consumption is concerned. The three-and-a-half-pound brain of a 170 pound human takes up about 20 percent of the body's total energy consumption. How much a dog's brain consumes is unknown, as far as I know. It's certainly lower than a human brain's consumption, since dogs don't think as much as we do, but it's probably also high. It's indicative that our dogs were totally exhausted after the thinking tests.

Different talents

In the following pages I'd like to talk about Wisla, the Saint Bernard female, Robby, the retriever, and Teddy, the German shepherd: three canine learning personalities with different talents.

INFO

Intelligence

When carrying out the learning and reasoning exercises you should remember that dogs – just like people – have different talents. Some dogs find it hard or impossible to figure out a task, while others can solve it in no time. But even dogs that are "learning resistant" are perfectly loveable and need your affection. Give them simpler tests and exercises to do. Take your pet's personality into account.

Wisla, hungry for knowledge

Let's start with Wisla, the Saint Bernard female. I've already mentioned that she was a year and a half old when she came to us. Her former owners loved her and taught her the fundamentals of basic training. Although she followed important commands such as "sit", "down", and "come", during our walks together she and I came in for quite a few surprises, in the truest sense of the word.

◆ **Fear of new things:** It was scary to see how little Wisla knew about the world around her. She was afraid of any new impressions. At the beginning, our walks were for her a total sensory overload. Again and again, she would sit and stare at traffic signs, advertising, and a lot of other things that were new to her.

I observed that the more she took in, the more cautious she became. It went so far that she refused to walk past a fluttering piece of cloth. She reacted according to the principle that offense is the best defense, and she started barking at the cloth. Her body language revealed her fear. Her tail was pulled in slightly and her head was bent down to the side. At this point I just had to step in.

◆ **Sensitivity is called for:** The first thing was to stop her from being afraid of the cloth. I took her by the leash, spoke to her soothingly ("good Wisla, good girl"), and we headed for the cloth. But just before we reached it she stood still and refused to budge. I pulled gently on the leash and coaxed her with sweet words into follow-

1

2

Something new for Wisla

The formerly distrustful, frightened Wisla has turned into an animal full of curiosity and a sense of adventure. Her self-esteem has increased enormously. There are only a few things now that can scare Wisla.

Wisla and Robby

Wisla hardly has problems with her surroundings anymore, while Robby, the retriever, reacts rather cautiously. But Robby's element is water. There, nobody can pull anything off on him. In water, he's the boss.

ing me. And she did follow me. When we were in front of the cloth I told her to sit, I petted her and continued talking to her. Using the voice is very important. It's soothing for the dog and it displays our feelings. We stayed put for a minute or two, and I let her sniff the cloth. I immediately repeated this procedure twice, in order to let her get used to this situation and gradually rid her of her fear.

Wisla needed to take this walk three times before she lost her fear of the cloth. Now, the sight of a cloth fluttering in the wind has absolutely no effect on her. She got the meaning: Fluttering cloths are not danger-ous. During our walks in the following three weeks I was careful not to let her be assailed by too many new things. She started with only small doses. The more she learns, the easier it is for her to handle new experiences. But it takes time. A whole new world is being created in her mind. Nerve cells are growing and forming new connections. (→ pg. 36).

◆ **Is Wisla really "chicken"?** Wisla was always good for a surprise. I was totally dismayed when I threw a stick into a stream for her to fetch and she backed away in fear of the waves. I can hardly describe what was going on in my head.

I was thinking, "I've ended up bringing a chicken into my home." But Wisla's quick learning ability and her endearing manner quickly drove that silly thought away.

In full view of her, I took the stick, put it on the water and told her to fetch it, making sure that there were hardly any waves in the water. Wisla fetched the stick. The second time around, I held the stick in the water and started splashing. Filled with curiosity, she took the stick in her mouth and started played with me. She got used to the splashing water. From here it was just a short step to getting her accustomed to the waves. First I dropped the stick from a small height into the water, and later I threw it far away. Wisla learned very quickly, yet I was puzzled as to why she was so afraid of the waves at first. A telephone call to her previous owner cleared up the mystery: As a puppy at the beach, she had been swept up by the surf and been scared badly. That's where her fear came from. Now she frolics joyfully in water. I could continue with a list of such experiences, but they all illustrate the same point: Wisla was frightened by things that were new and surprising.

◆ **Caution is not the same as fear:** Her behavior may seem annoying to us, but from a biological point of view it makes sense. Her reactions are like those of nearly all wild animals. Anything new is approached with caution. After all, it could be a life-threatening danger. I could take away Wisla's fear by calming her with my voice and making the situation familiar to her. As a rule, I let her sniff at anything new. Wisla is not an animal who is frightened from birth – that much was clear from her behavior. Her fear may have been great, but with empathy and support her curiosity overcame it. The fact that she learned quickly to handle the new situation, and the way she reacts nowadays to new things, also says a lot about her abilities.

◆ **Her self-confidence grows:** There are only a few things left which can scare Wisla. Her self-confidence has grown. This has big advantages for her owner. A dog that is self-assured behaves more reliably. It doesn't become aggressive out of fear and it's less prone to biting. In short, it's less dangerous to other dogs and to humans.

But why was Wisla so frightened in the first place? It's quite simple: She had had too little interaction with her environment.

Learning to use a food bowl at the tender age of four weeks takes learning. But it won' t take long.

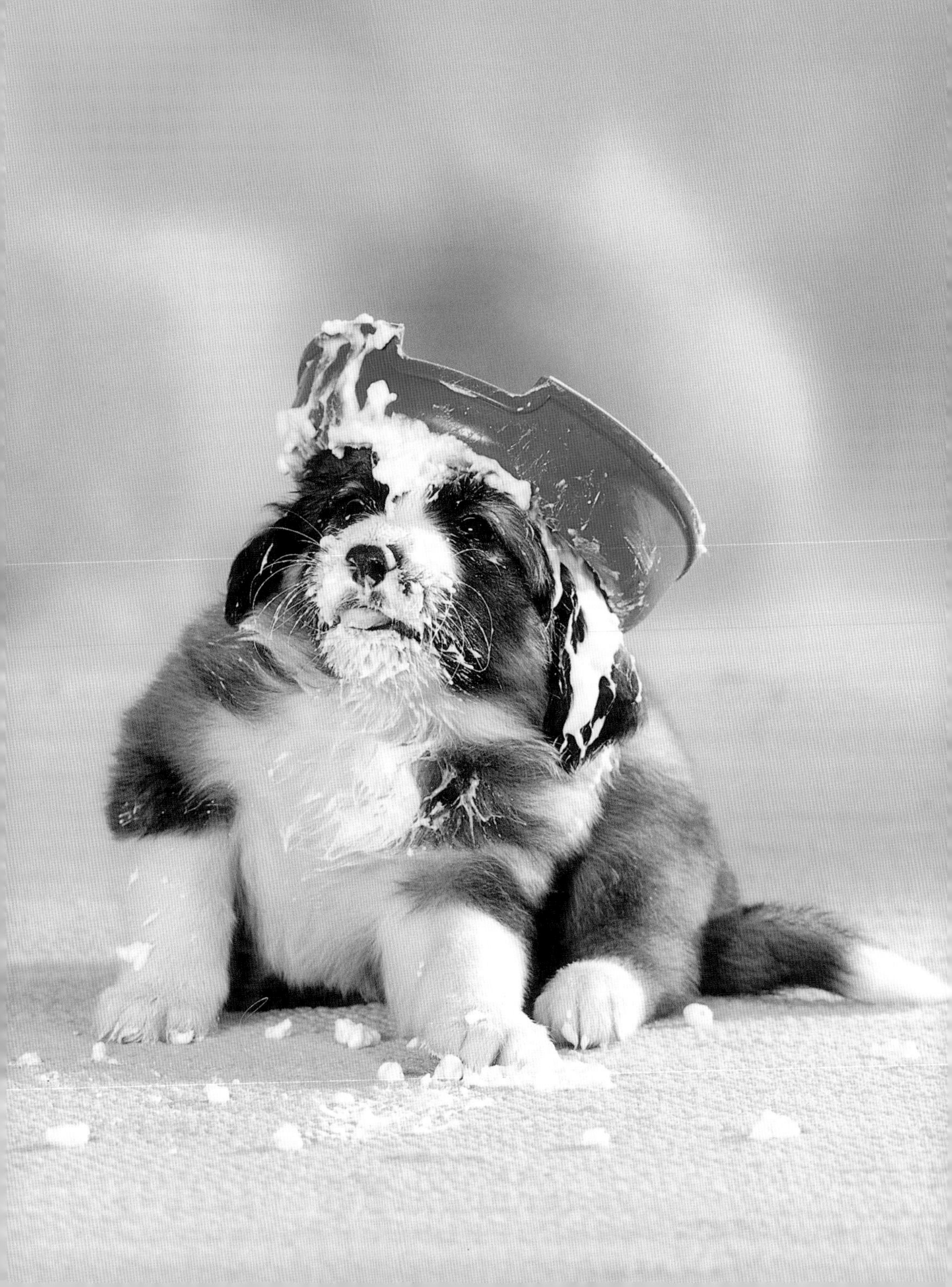

◆ **Experience is important:** Unlike Wisla, other puppies were able to play more often with their fellow dogs, and they had contact with the world of humans. They learned how to move in street traffic. At an early age, they learned that there's nothing to fear from a traffic light, a screeching car, or a cyclist. Wisla, on the other hand, was raised in idyllic surroundings. She had her backyard and the love of her owners, two very important things, of course – the foundation for healthy behavioral development – but nonetheless not enough to keep her mind sufficiently busy.

◆ **Dogs can, and need, to do more:** Wisla demonstrates this every day. She observes her surroundings carefully. The following example shows just how carefully. After going for a walk, we always clean the paws of our two dogs. First comes Robby, the retriever, then Wisla. While I'm cleaning Robby's paws she waits patiently in the car, with the doors open. Without my having taught her, she comes trotting to me as soon as I'm finished with Robby's last paw. On her own, she's learned to recognize when Robby's cleaning is done, and when it's her turn. She also quickly noticed that she has to raise her paws for me to wipe them. Cheerfully, without a word from me, she stretches out one paw after the other. This eagerness to please is simply irresistible. Her reward is gentle hugging and petting, which makes us both feel good.

You can teach a "big eater" like a Saint Bernard all kinds of things by using food as a reward. In no time at all, Wisla learned that she had to stop at the curb and not cross the street until she heard the

Robby loves water more than anything else. He feels safer here than on land.

command "go". Our big dog has an agile mind, as you'll read more about later in the book.

◆ **Learning is easier in the early years:** Fortunately, Wisla was still relatively young, otherwise it would have been more difficult to help her lose her fear and bring out her curiosity. She was still playful, and it was through her playfulness that she conquered her environment. This "neuronal" window gradually grows smaller with increasing age. And this applies not only to dogs. Humans also learn certain things better in early childhood. A good example of this is acquiring a foreign language, something we as children learn with great ease, but which as adults we have to slog through with great effort. To rephrase the old saying: "It's hard to teach an old dog new tricks".

◆ **What's important for you and your dog:** During puppyhood and adolescence dogs learn a great deal in a short time. During this phase their neuronal window is especially wide open to particular learning processes. A dog learns to trust its human partner. Dogs that grow up without contact to people almost never form relationships with humans. A young dog is supposed to learn a lot – a lack of stimulus is harmful. Learning has a positive effect

INFO

Learning in later years

Older dogs are also good at learning, and they enjoy it. Of course, you have to take into consideration the physical fitness of your old-timer. If there are any doubts, take your dog to the vet. Don't practice too long with your pet and give him longer breaks. Even an older dog with a lot of experience will enjoy learning new things. That's why you shouldn't forget to regularly encourage your old friend both physically and mentally.

on its self-confidence. A confident dog is usually easier to handle. Gaps in training can be compensated for only through patience and understanding. Punishment has a destructive effect. It destroys not only the trust that's been built up, but also the dog's self-confidence.

Robby, the scaredy-cat

Our second candidate is Robby, a retriever and a totally different type of learner from Wisla, our Saint Bernard.

Robby was a puppy when he became part of our family. Being the cutie-pie that he was, he was pampered and spoiled by the children and his owners. There was no lack of love and affection for him. He grew up the same way as thousands of family dogs do. Nothing special about that, except perhaps he wasn't trained quite thoroughly enough. Still, he learned everything a family dog is supposed to learn. It wasn't always easy, since he could be thick-headed, but that wasn't the problem.

◆ **Can fear be overcome?** As he got older, Robby developed certain fears. As an adult dog he was afraid of other, much smaller dogs. He was cautious and mistrustful towards anything new. Again and again I would approach him gently with the same object, but it still scared him. A good example is a scenario I set up every day: There's a newspaper lying on the stairs. Robby goes up to the paper, stops and waits. It would be no problem simply to walk past it, the way Wisla and Teddy do when they go ahead of him, but Robby can't manage it. (→ text, pg. 46).

Playful training

Training should be fun, and it shouldn't turn the dog into a spineless order taker. For this, you have to know exactly what your dog's strengths are.

■ Be a good judge of abilities

A good dog trainer first observes his student carefully in order to judge his abilities. This is always the starting point for successful training. Young dogs reveal their strengths when they play. Interaction with other dogs shows if it's bold or aggressive, cautious or even fearful. But that's not all. The way a dog investigates his environment when playing says a lot about its curiosity and mental abilities. Dogs that don't show much interest in things are later difficult to motivate during training. A good dog trainer recognizes these abilities and knows how to encourage and challenge his animals. With this knowledge in his head, he's ready to begin training.

■ Why training is important

There's no doubt about it: Dogs have to be trained, for the sake of their owners and themselves. Only dogs that have gone through basic training can later take the reasoning tests. It's relatively easy to teach a dog its ABC's, since dogs are pack animals. A dog can survive as part of the pack only by paying attention to the signals that others give. It's also reassuring to think that dogs have been bred throughout the centuries to understand our signals. But just what are the ABCs of basic training? I think that dogs should master the following commands: "sit", "down", "let go", "stop", "come", and "heel". For some this may not be enough, but as soon as your pet has them down pat, life becomes a lot easier. The tools you use in training are your voice, your facial expressions, and your hand and arm gestures, as well as with the right rewards.

■ A few training tips

The "stop" command: How can I teach my dog to obey the "stop" command in order to keep it from running away from me? A secret of good training is to be able to anticipate within a fraction of a second what your dog intends to do. This becomes easier the more bonded you are with your dog. When you sense your dog is about to start running, give him a short, sharp "stop" command. The dog will be "stunned" that you've read his mind, and as a reflex he'll turn to look at you. This is when you give him a reward and say "good boy" using a drawn-out, gratifying voice. Repeat the situation several times until the dog has understood the word "stop". When he's got it down pat, put him in a tempting situation. Take him to where he likes to run and play – usually somewhere where there are other dogs. Here you'll have to be a step ahead of your dog. At that moment where he sees another dog, give the "stop" command and continue as described above. If your dog tries to run for it, grab him by the collar and with a firm voice say "stop". Your voice has to convey determination to the dog. If he reacts the way you want him to, praise him and give him a treat.

Education, dog-style: Baby is chewing on Mama's ear. At first she's pretty calm about it.

But when it gets to be too much, she takes Baby's muzzle in her jaws. "That's enough!"

The "come" command: Your dog has become immersed in the odors of a meadow. You call out "come" desperately, but he doesn't hear you. What can you do? I've learned a simple but effective trick from watching the great animal trainer, Gerd Semonit. It works on lions, tigers and leopards, and it works on dogs, too. When going for a long walk, take along some paper that crackles easily. If your dog won't come to you, get within 15 to 30 feet of him without calling, then crumple the paper. The noise will immediately attract his attention to you. This is the moment when you say "come" in a sharp tone, waving a treat in your hand. With this small trick you're bringing your pet back into your world – without punishing him. Try it out!

◆ **Robby knows how to get help:** Even the tastiest treat couldn't get Robby to go up the stairs past the newspaper. All my encouraging words were useless. He was rooted to the spot in front of the paper. Now you might think that he's just plain stupid. But that's not the case, because he's actually developed his own strategy: Robby has trained us. He starts barking whenever he can't, or doesn't know, what to do.

It took years for him to learn how to use his muzzle to open a door that's slightly ajar. It was also impossible to teach him to fetch a stick or a glove, although occasionally he would proudly carry a stick on his own accord. All of my training skills failed to teach him to bring me anything. But that's not all. It's almost impossible to get him to go to a chosen spot. To give an example, I'll point with my outstretched arm and index finger to a spot in the room where he's supposed to go. No way – no way, you have to lead him there and give the "down" command. Then everything's okay with him. This was never a problem with any of my other dogs. What's the matter with Robby?

◆ **A gentle push for his own good:** It's striking how carefully Robby has to be handled during any type of training. Even using the most tempting reward I can hardly get him to do anything. He refuses to budge at the slightest push. And yet, he sometimes has to be pushed for his own good. A good example of this is getting into the car. The VW model that we drive opens up in the back, where it's easy for the dogs to jump in. It's no problem for the other dogs, but for Robby it was.

Even though I had put his favorite treat on the floor of the car directly in front of him, he wouldn't jump in. All my tricks and efforts were in vain. The only thing that helped was to gently force him. I took his front paws, placed them on the floor of the car, and lifted his rear into the car. All this time, I spoke soothing words to him.

TIP

Recognizing a dog's skills

Dogs often display their particular skills when they're playing. One dog will be physically very gifted, while another enthusiastically uses his nose and other senses to explore new surroundings. You should encourage individual skills. You shouldn't, for example, make your Saint Bernard stand on his hind legs, something which for a terrier is a piece of cake. The Saint Bernard, however, might be very eager to solve mentally challenging tasks.

The sheer joy of living. Energetic playing feels so good after difficult mental tests.

I repeated this procedure four times before he jumped in on his own. Robby has learned many things in this way. You might think that Robby isn't particularly intelligent, that he's simply a slow learner, but you wouldn't think that if you saw how Robby behaves in water.

◆ **His element is water:** Robby is a water baby, and water is his elixir of life. This is where he feels safe and sound. He's passionate about fetching sticks in the water and defends them against bigger dogs such as Teddy, the German shepherd. So it's simply not the case that Robby is stupid. Yet on land other factors such as fear are more powerful. I don't know why that's so, and I can only speculate.

He had a proper upbringing – that is, a happy puppyhood – so that can't be the cause. We have no way of knowing what happened to him in his first weeks of life. We love him with all his faults and weaknesses, and we accept his personality. Robby's case is a perfect example of recognizing that you can make an animal do only that which it's capable of doing. Learning ability is determined by genetic substance or through early bad experiences.

◆ **What's important for you and your dog:** It's difficult to exactly judge a dog's learning skills. Many skills remain hidden from owners, since they unfortunately don't have the key to discovering their dog's talents. What's certain is that dogs have different areas of ability, and that for some dogs learning is easy, while for others it's difficult, just the way it is with us humans.

1

Teddy springs into action

There's no stopping Teddy, the long-haired German shepherd. Physical activity is as important to him as mental stimulus. Out of all my dogs, he was the most gifted learner, but also the most demanding one.

Teddy, the overachiever

Teddy came to me as a cuddly long-haired German shepherd puppy. He remained curious even at an advanced age, and nothing escaped his nose, eyes and ears. He was always in the mood to play. Unlike Robby or Wisla, during walks he could hardly keep busy by himself. He would always challenge me to play with him.

◆ **A fast learner:** Teddy had the qualities that made him an ideal dog to teach, and he challenged me to teach him. Nothing was safe from him. On his own, he learned

2

3

Never a dull moment

With a partner, Teddy doesn't know the meaning of boredom. He always wants to play with someone. Sometimes he'll wait until he's given a task, which he usually carries out with zest. It can be very exhausting, however.

The pleasure of playing

Teddy in action. When going after a ball, his movements are almost acrobatic. The only thing that's a bit tricky is obeying. As a rule, Teddy ignores a first command. He needs to be treated lovingly, but firmly.

how to use his paws and muzzle to open food packaging quite skillfully. It wasn't easy to cure him of that habit. Handling him required firmness and persistence. If I said "let go" in a friendly way, it was ignored. Only a forceful voice and a commanding stance had any effect. Keeping Teddy busy was physically and mentally exhausting. I continuously had to think up new ways to satisfy his eagerness to learn.

◆ **An excellent memory:** Teddy's good memory surprised me. I taught him to recognize the difference between the marked lids of two food bowls. One lid

was marked with circles, and the other one with squares, but only the lid with the circles had food under it. After four attempts he knew where I had put the food. More than one year later I did the same experiment with him – right away he chose the right bowl. Experiments such as this are a key to understanding your dog's learning abilities (→ mental workout, starting on pg. 68). In the end, both dog and human benefit from them. The human discovers what's going on in his dog's mind, and the dog has fun and stays mentally fit.

How does a dog see itself?

What does a dog know about itself? Have you ever asked yourself this question? Here you can read about our findings up to now.

■ Do dogs know what they're doing?

Does Robby, the retriever, have an idea of what he's doing when he swims after a duck? Is he aware of himself, or is he in some way behaving blindly and unconsciously? I'm pretty sure that he sometimes realizes what his actions are, but scientific proof for this is limited to rats. In experiments, rats had to press a button when they wanted to eat. There were, however, four different "behavior" buttons. If the rat first cleaned itself and then wanted to eat, it had to press the "clean" button. If it rested before it wanted to eat, it pressed the "rest" button, and so on. It was possible this way to determine that rats do indeed know what they're doing.

■ Do dogs know about their bodies?

Do dogs have a concept of their own bodies? I believe so, since this subject has been thoroughly researched in other species. A stag, for example, knows about the splendor of its antlers. It also knows if it's carrying its antlers or not. Knowing about one's own body is anything but obvious. A small disturbance in the brain is enough to deactivate this element of the conscience. I've seen how leopards will eat their own tails following an injury. They aren't able to distinguish between what's theirs and what's foreign. Such self-mutilation usually has fatal results. Dogs certainly are conscious of their bodies, and it helps them to learn something about themselves.

■ Do dogs know who they are?

What about dogs and their sense of self? Let me mention first that dogs don't pass the classic mirror test, which is used to see if animals know who they are. What's the mirror test? In this simple, but very revealing test, you put an animal in front of a mirror and observe its reaction. Most animals see in their reflection a member of their own species which they then either attack or court. I myself have either carried out the test or witnessed how animals of highly different species react. All the dogs I've tested have either barked at their reflection or bared their teeth. Apparently they thought they were looking at another dog. It's interesting that even timid dogs did not pull in their tails or exhibit behavior associated with fear. Fear thus did not emanate from their own reflection. It seems that dogs don't possess an ego, a sense of self, but I'm not quite sure. Perhaps they define their sense of self through smell.

■ Who recognizes himself in the mirror?

When I was allowed to hold a mirror in front of Xindra, a female chimpanzee at the Basel Zoo, I received a revelation. She examined every part of her body she could see in the mirror. For minutes, she picked at her teeth. She examined one tooth at a time. Then came her lower body. She used her fingers to spread

Does the kitten see an enemy in the big dog? And does the dog know that's a cat in front of him?

open her anus and vagina and she peered inside. After having seen this, you really don't need scientific proof. Xindra's keeper dabbed a small spot of paint on her forehead without her noticing. It didn't bother her, so she didn't rub it off. Then we showed her the mirror. She was baffled when she looked at herself, noticed something wasn't right, and rubbed the spot off her forehead. That was definite proof. Why? Children younger than a year and a half or two will try to grab the spot in the mirror because they haven't yet recognized that they're the ones who have the spot on them. They don't realize this until they're over two years old, at which age they will wipe the spot of their forehead. In humans, self-awareness is formed at about the age of two.

Do dogs have a sense of self? Unfortunately, the mirror test doesn't give a definite answer.

Recognizing skills and helping develop them

A dog should be trained according to its temperament and skills in order to fully develop its personality. For this, you need insight into your pet.

My dogs Wisla, Robby, and Teddy clearly demonstrate that they're equipped with different skills. Each one of these three canine personalities has to be encouraged in a different way.

Training Teddy, the German shepherd, calls for a firm hand. Robby, the retriever, needs to be handled gently, and big Wisla requires sensitivity and understanding.

Since the eagerness to learn varies from dog to dog, I can't demand that each one learn the same things. That may sound obvious, but in dealing with dogs it's often easy to forget it. There are indeed typical characteristics for the German shepherd, retriever and Saint Bernard breeds, but they form too general a picture to enable us to understand what's going on in the mind of a particular dog. We have to recognize Wisla, Robby, and Teddy as individuals. Simple learning and reasoning tests can help us in our search (→ pg. 68 onward).

Can dogs count?

Counting is one of the things we take most for granted in the world. But is it really so? There are tribes in the Amazon region that count only to four. Five or ten pieces of fruit are simply too many. They don't make any further distinctions. Why are we so skilled with numbers, while others aren't? It's a difficult question to answer, but one thing is for sure: In our technological world we can't exist without numbers. To attribute a definite number to an amount of objects requires a certain ability to think in abstract terms.

The ability to count enables us to play out certain logical operations in our mind. It's no wonder, then, that mankind has always wondered if animals are able to count. "Clever Hans" was a horse that became famous by stomping its hoof to show its

arithmetical skills. Five beats of the hoof showed that two and three equals five. It could even figure out the square root of 16: Without hesitating, it stomped four times with its hoof. Was this an animal genius or a hoax? This question was on people's minds in the early decades of the last century. Understandably, "Clever Hans" became a challenge to the scientific community. It took two committees of experts to solve the riddle. A young staff member of the Berlin Psychological Institute was able to prove through an exhaustive series of experiments that whenever the horse tapped out the correct number, spectators would give a "twitch of relief" – an unnoticeable lifting of the head. This tiny movement – sometimes not more than a fraction of an inch – occurred involuntarily and unconsciously. But the horse took this as a signal to stop tapping. Unnoticed by its teacher, it had been trained to do this during the course of the lesson

What's a horse doing in a book about dogs? I've given this example to show you how careful one has to be when planning an

How often can your dog correctly choose the number that hides the food?

experiment in order to get a definite result. It's too easy for a mistake to slip in. With this difficulty in mind, we tested dogs according to a method we had developed for cats. Some of our cats were able to count to four in the mathematical sense. The principle behind the experiment is simple.

The cat walks up to four food bowls placed about four yards away. The lids have one, two, three or four symbols on them, and one, two, three or four notes are played. The cats gets food only if it chooses the bowl whose lid has the same number of symbols as the number of notes that are played. It would be going into too much detail to describe all the control experiments, but let me just say that the cat heard three notes the first time and chose the three symbols. With four notes there were also no mistakes. It headed straight for the right bowl. These correct decisions show that learning was not involved. Of course, we checked our results statistically.

◆ **Dogs can count to three:** We were convinced that we could test a dog's counting skills using the same experiment we used for cats. Confident of success, we set to the task of obtaining the evidence. Unfortunately, it was in vain. When the dogs started having difficulty deciding between two and three, they either lay down between the food bowls or they looked to their owners for help. We attempted everything possible, but after a year and a half we gave up dejectedly. This, however, doesn't mean that dogs are incapable of counting. It could also be the case that when the situation starts getting difficult

This clever border collie just seems to be waiting for new and interesting tasks.

they leave the decision making to their owners, even though they understand what's happening.

This idea has been reinforced by experiments done by scientists at the University of Budapest. They presented some dogs with the task of winning a treat. The dogs had to use a lever to release the food, which was within view. Some of the dogs quickly figured out what to do, while others remained passive. What was striking was that the dogs that had an especially close bond to humans were extremely passive.

This passivity immediately disappeared when their owners encouraged them to get the food. Then all the dogs knew what to do. This doesn't mean that dogs that relate closely to humans are "dumber" – it only indicates dependent behavior. Perhaps this was also the reason we didn't have any luck with our attempts at counting. Rebecca West and Robert Young were luckier. They used a very different approach. Their method achieved success. Their results lead to the conclusion that dogs can count. They were able to add 1 + 1 and 1 + 2. Much work remains, however, before we have a definitive answer.

INFO

Dreams

It has been established that in dogs, just as in humans, the phases of sleep alter with increasing age. REM phases (→ pg. 58) decrease with age. Older dogs also apparently dream less frequently. You can observe this in your own dog. I happen to own a young dog and an older dog, and I can see that the young dog moves its legs much more in its sleep than the older one does.

Rico, the superstar

There's no doubt about it. Rico, a border collie belonging to Ms. Baus, is a star. Who else has managed to be a guest star on one of Europe's biggest TV shows, as well as appear in "Science", one of the most renowned natural science publications. What's so special about Rico, and what can the guy do?

◆ **What Rico can do:** On command, Rico is able to pick out the correct stuffed toy out of 100 stuffed toys and bring it to his owner. He'll go straight to the teddy bear, the stuffed sun, the Dortmund team soccer ball, or the Munich team soccer ball. Learning the names of 100 different objects is an enormous task. You can see for yourself just how enormous it is. Just try memorizing 100 Chinese words. You'll see how hard it is. The Chinese language is probably as foreign and unfamiliar to you as our language is to Rico. Rico amazed both the TV viewing public and the scientific world. At first they thought that perhaps this was another case of the "Clever Hans effect" – basically, that Rico was a hoax (→ pg. 53). Juliane Kaminski, a researcher at the Max Planck Institute in Leipzig, looked into the matter and allowed us to attend and film her investigations.

Observing Rico was quite an experience for me as a biologist. First, Ms. Kaminski picked 15 stuffed animals at random and placed them around the room. Rico's owner, Ms. Baus, was in another room during this time and thus couldn't see which animals Ms. Kaminski picked and where she placed them.

◆ **Getting proof of reasoning:** Ms. Baus called Rico to her and gave him the command: "Rico, fetch the butterfly." What then happened surprised me. Rico quickly ran to the other room, but once he was there he then carefully went from one stuffed animal to the next, sniffing at them. This didn't look like simple, blind memorization, but more like deliberation.

Two years later, Ms. Kaminski was able to further confirm this through an elegant experiment. Under his familiar toys she placed a new, unknown toy, and called out the name "rooster". Rico had neither seen the toy nor heard the name before, but still he could correctly pick it out and grab it.

INFO

Emotional intelligence – what's that?

Dogs are adept at analyzing and interpreting a human's emotions. In a close human-dog relationship the dog even responds to the feelings of his human partner. He can handle the human's moods. A bad mood, for example, will make him withdraw, and grief makes him want to snuggle up and comfort a human. He reacts to joy with high spirits and playfulness.

This was convincing proof that dogs can think logically. Rico had reasoned according to the rule of exclusion: the new word must describe the unknown object. Small children learn new words according to the same principle. This caused a scientific sensation and earned publication in the scientific journal "Science".

But there's more to Rico than his sunny side. Handling him is not easy and he requires a lot of energy from his owner. His "addiction" to learning can hardly be stopped. He's always dragging toys around, and he keeps his owner on her toes. But what he hasn't learned yet is how to deal with his fear of thunderstorms. At the first distant rumble of thunder we had to pack up our cameras, since he had crawled away and wouldn't let himself be seen.

◆ **Recognizing talent:** How did Ms. Baus actually recognize Rico's talent and his gifts? As so often happens in life, it was through coincidence.

Rico had hurt himself and wasn't allowed to go on long walks, so she had to think of something to occupy this bundle of nervous energy. That's how she got the idea of his fetching the stuffed animals. And it was a great idea, I must say. It exercises his memory and his muscles, and the dog constantly learns new names for individual stuffed animals (→ pg. 55). This is physical, and at the same time, mental training. It's good for us, too, to keep our memory fit by learning new vocabulary.

As a dog lover and scientist, I was of course overjoyed to see the scientific proof that dogs can think logically to solve problems.

Dogs are excellent at recognizing human emotions. They thus possess a high degree of social intelligence.

Emotional intelligence

In some respects dogs are the most accomplished animals, outdoing even chimpanzees, which are considered the most intelligent animals on the planet.

Dogs possess a great degree of emotional intelligence (→ opposite), which gives them a special relationship to humans. I don't think there's another species that understands humans so well, and that's able to recognize and respond to our feelings as unerringly as dogs can. This is why people often feel especially close to their dogs. From the beginning, dogs have been put to use in the service of humans. For humans, the aim of raising dogs was to obtain useful traits through selective breeding. Unintentionally and unconsciously, breeding also promoted the social and emotional intelligence of dogs – as a by-product, so to speak. This developed into a cycle of mutual dependence.

The more dogs understood us, the stronger their ties to us became. We thus raised dogs that displayed the characteristics we were looking for, and we continued breeding them selectively.

Sleeping and dreaming

Healthy sleep is as important for dogs as it is for us. Our pets apparently use dreams to deal with the events of everyday life.

■ Sleeping and thinking

Thinking and sleeping are inextricably bound together. This opinion is also maintained by J. Allan Hobson, a professor at Harvard University. We know how lack of sleep affects humans: We can't concentrate, there's a lack of attention, and we can barely perform actions that require coordination, such as driving. Why should this be different in dogs? Of course, a dog's sleep is different from a human's. Hobson argues that a more complex brain means more complex sleep. What does the brain do while we're asleep? Is it also asleep? What's clear is that the brain is active during sleep, but it doesn't process sensory perceptions. Brainwaves are measured during sleep (through an EEG = electroencephalogram), but they're quite different from the brainwaves in a waking state. Sleep is a result of brain activity and is thus a useful function. Brainwaves allow us to distinguish among different stages of sleep that are periodically repeated. One stage is called the REM phase (rapid eye movement), which as the name describes, is distinguished by rapid movement of the eyeballs. During this stage there is a great deal of dreaming, and individual areas of the brain are particularly active. In addition, animals that have gone through training programs experience an increase – however small – of REM sleep. Sleep deprivation makes additional learning difficult. During REM sleep we reinforce what we've learned, but we're not behaving. Instead, we recall memories and play them out in our dreams. This is according to J. Allen Hobson's theory. What's interesting for us are the effects of sleep on thinking. We now know that the information we obtain in our waking hours is saved, actively updated, and processed while we're asleep.

■ Dogs also dream

There's every reason to believe that dogs dream. As with us humans, similar brainwaves can be measured and high neuronal activity can also be observed during REM phases. Dogs' biochemical mechanisms are also very similar to those of humans. There are hardly any doubts that dogs dream. You can observe this yourself in your pet. I own a young dog and an older dog and I notice how much more often the young dog twitches its leg during sleep. This makes senses, since for a young dog the world is new, and a lot of what it experiences has to be digested. This is accomplished apparently in their dreams. The difficult question is, what are they dreaming about?

Adrian Morrison, a famous sleep researcher in the veterinary department of the University of Pennsylvania, showed us a striking example of what a cat dreams of. The cat in question had a tumor that took away the inhibition that would have normally kept it from using its running muscles during sleep. This inhibition is likewise missing in sleepwalkers. That's what allows them to walk in their sleep. But back to the cat. Adrian Morrison had

This puppy is relaxing comfortably in a hammock. Will it have sweet dreams later when it's asleep?

taped how this cat had gone hunting for mice while it was dreaming. It stood up – its eyelids open, its nicitating membranes (inner eyelids) closed –, crept along the floor and suddenly sprang onto its supposed prey. This was a hunting scene as it would appear in waking life, but the cat was asleep, as evidenced by its closed inner eyelids. The majority of researchers agree: animals that possess a well-developed cerebrum also dream. And it's no coincidence. They are the thinkers of the animal world.

■ What's important

A well rested dog is better than a tired dog at solving problems. You should remember this when you do the intelligence tests with your dog (→ pg. 71). Take his sleeping patterns into account.

After some vigorous playing, this puppy has fallen asleep on its mother's back.

How well do dogs understand us?

Work in this field of research is being carried out particularly in the department of behavioral science at the University of Budapest, under the direction of Adam Miklosi at the Max Planck Institute (MPI) in Leipzig. Miklosi found out that dogs understand many human gestures. Even when its master is pointing at a distant object, a dog will understand that it's supposed to run to it and fetch it. That's nothing special, you may think. But you'll understand how difficult it is when tests are done on other species. Just how far do humans and dogs understand each other? Do dogs also understand indirect suggestions from humans? Brian O'Hare at the MPI in Leipzig tried to find out, with satisfactory results. We did the same experiment in slightly modified form, and you too can easily carry it out on your dog (→ "recognizing signals" and "the pointing test", pg. 77).

Puppies that are raised by a cat will also later in life clean themselves like a cat.

Can dogs cheat?

Can dogs figure out what we're seeing through our eyes? It seems so. The experiment again took place at the Max Planck Institute, and again we copied it. The test is interesting and at the same time fun. The only requirement is that the dog must be good at obeying. The subject in our experiment was named Gina, a big, clever dog. She was a cross between a Saint Bernard and a Bernese mountain dog. Gina's owner sat down comfortably in a chair, and Gina lay down a short distance from her. In front of both of them, about three or four

yards away, there was a snack lying on the floor. Gina had received the "down" command, and she was not to touch the treat. Every attempt to get near it earned her a sharp "stop" from her mistress. And so they sat there for three or four minutes, with the rules laid down. Gina obeyed. Now began the actual experiment. The owner picked up a newspaper and opened it in front of her face. She made believe she was reading. And what did Gina do? She paid careful attention to her owner. Repeatedly, she looked first at her mistress, then looked at the treat. This game was repeated at least ten times. Her mouth started watering. She began to stand up and then sit down again without moving from her spot. This procedure was also repeated several times. In the meantime, nine minutes had gone by. Suddenly, Gina stood up and walked – more like tiptoed – towards the treat. She gulped it down, then sat down somewhere else in the room. Gina wasn't smart enough to go back to where she had been sitting. It was clear to us that Gina had a bad conscience. In the control experiment – during which the owner sat silently, but without a newspaper in front of her face – Gina did not cheat. They both sat next to each other for 45 minutes without a mishap. We tested three other dogs and got the same results (→ "the cheating test", pg. 79).

INFO

Contact is important

It's extremely important for dogs, being pack animals, to have plenty of contact with humans and other dogs. Otherwise, they won't develop their social skills. Dogs that live permanently isolated in cages have difficulty relating to other dogs and humans. They become aggressive and they bite. Keeping a dog this way is a sure recipe for creating a fighting dog. Most of these unfortunate creatures are thus the victims of man.

Can dogs imitate others?

Imitation does not hold high esteem in our society. A person who copies doesn't possess creativity or originality, according to general opinion, and there's some truth to that. But I think this point of view falls too short, since there's more to imitation than that. By watching and imitating, you spare yourself the trouble of learning by trial and error. And you don't have to act out and judge problems in your mind. Instead, you see the solution right before your eyes. You only have to copy it. But what does "only" mean? Imitation requires that you first understand what the other person is doing, and second, that you're able to adapt the other's way of doing things to your way of doing things. You're adjusting yourself to the standards of others. There's a problem inherent in imitation: The behavior being imitated rarely occurs alone. Instead, it's interwoven with other types of behavior. It's hard to draw a clean line between particular types of behavior. When a lion, for example, brings down an antelope, part of this behavior is inborn, part has been copied from watching others, and part has been learned by the lion on its own.

◆ **Learning through imitating:** This problem was recognized by no lesser a personage than Charles Darwin, the creator of the theory of evolution. He astutely described several cases of puppies being raised by cats where we see how puppies became accustomed to typical feline behavior. They licked their paws to use them like washcloths on their heads and

1

How can I express what I want?

The cell phone is on a shelf out of the dog's reach. When it starts to ring, he wants to bring it to his disabled owner. But how can he do it? In this scene we've reenacted how Philipp solved the problem.

2

The dog figures it out

Philipp brought an empty film container to a friend who was visiting Richard in order to get her attention. The container, which Philipp had never used before, was the equivalent of his saying, "I need help."

ears. One of the puppies kept up this habit throughout the thirteen years of its life. This was in 1871. Darwin recognized a dog's mental skills. It's all the more astonishing, then, that after more than 100 years we still underestimate our dogs and insist on rote training. Let's take as an example the training that drug detection dogs undergo. Usually the trainer receives the puppies when they're eight weeks old. At three months, they receive their first training lesson. Then the more gifted puppies are selected. More training follows and at the end there's a final exam. It's a tough year of schooling, but actually most of it could be spared. At least that's according to a study made in South Africa. Puppies born to a drug detecting female were allowed to participate in her tracking work instead of being sent to training. The results were astonishing: Learning at their mother's side was just as effective as the tough training with humans. The young dogs had indirectly learned their mother's work, and then imitated it with playful ease.

Up to now, the subject of imitation has hardly attracted attention. This may be due to the fact that there's usually only one dog in a household. I've noticed, however, that

3

The guest followed the dog and knew right away what he wanted. He took the cellphone and brought it to his friend Richard. You could see that Philipp was satisfied. From then on, the empty film container meant "help me".

our two dogs have imitated the behavior of others. For ten years our retriever, Robby, was unable to push a door open with his muzzle until he saw Wisla, the Saint Bernard, do it front of him. During all the previous years he would stand helpless in front of the door barking for us to help him. It's interesting that he didn't learn this behavior from Teddy, the German shepherd, with whom he had also lived for years. This fact naturally makes a scientific approach more difficult. Is it possible that dogs have a certain partner who they'd rather learn from? In Robby's case, I guess so, since he copied other things from

Wisla. And Wisla? I think so too, although in her case it's not so clear. I'll have to observe her more closely and develop the appropriate tests. In Hungary, Adam Miklosi has already done this. His subjects are Philipp, a Belgian shepherd dog, and Richard, Philipp's owner. We'll be meeting them next. Richard is disabled and uses a wheelchair. When Richard lifts the front wheels of his chair from the floor to tilt it backwards slightly, and then brings them back down, Philipp imitates him by lifting and then lowering his front legs. But it's gets better: When Richard pivots to the left using his rear wheels, Philipp also turns to the left. The same goes of course for the opposite direction. We saw it with our own eyes and were stunned. What began as a game turned into science.

Can dogs understand symbols?

We've already met Richard and Philipp, a "dream team" as far as dog-human relationships go. Philipp holds the same position in Hungary as Rico does in Germany. Both dogs are unbeatable. In 2000, Philipp was voted the most intelligent dog of the year. Visitors to Richard and Philipp get their first surprise at the front gate. After we rang the bell, a boisterous shepherd dog came running up to us, barking. The gate was locked, and from a distance Richard called, "It's okay, they can come in." Philipp ran into the house, grabbed the correct key from the key rack, and gave it to the visitors. Not bad, but this is still on the learning level, and perhaps it's nothing special for a dog trained to help the disabled. But what comes next I've seen only in chimpanzees.

◆ **Philipp "talks" to his owner:** Philipp can also express his own wishes. He uses symbols to do this. They hang right next to the key rack. A triangle on a chain means "I want to play". A ring on a chain means "I'm thirsty". A chain with a cord means "I'm tired", and one with a plastic sausage means "I want to go for a walk". But there's one symbol, an old film container, that has a special history behind it. One day, a friend of Richard's had put Richard's cellphone on a shelf in a kitchen cabinet without thinking, thus placing it out of reach of both Richard and Philipp. An hour later – after his friend had left, and a new visitor had arrived – the cellphone rang. Philipp ran to the kitchen cabinet to fetch the phone, but of course he couldn't reach it. Since Richard and his visitor were sitting in another room they didn't notice Philipp's problem. But Philipp didn't give up yet. He quickly grabbed an empty film container lying on the kitchen table and took it to the visitor. This was his way of attracting his attention and getting him to follow him into the kitchen. The visitor did so and immediately understood what Philipp wanted from him. He picked up the cellphone and took it to Richard. Philipp was now satisfied. The film container became a new symbol that now means "help me".

He has also used the film container in similar situations when he didn't know what to do. This marks the beginning of a sign language. Each object or symbol corresponds to a word or a phrase. Deaf mutes communicate with each other in this same way.

Science has employed this method in investigating the use of language in apes. The greatest genius among apes is Kanzi, a bonobo. His sign language includes over 100 symbols, and he can even make sentences with them. I witnessed his skills, and I must admit it was a great moment in my life. Kanzi and I were playing with a ball when suddenly he ran to his laptop, where the symbols are displayed on the screen, and carefully pressed on a few symbols. They meant "pet me", as his trainer explained to me. This was just one convincing example of his skill. I'm convinced that Kanzi understands our language to a certain extent and that he can also use it.

I wouldn't go so far as to say the same thing about dogs, although Philipp also surprised me. When he became tired during our filming, he fetched the chain with the cord on it, the symbol for "tired". And when he didn't feel like continuing at all, he brought the chain with the plastic sausage, the symbol for "take a walk". There was no mistaking the message, and we stopped filming.

Learning Ricostyle (→ pg. 78).
Try to teach your dog to bring
you a toy by its particular name.

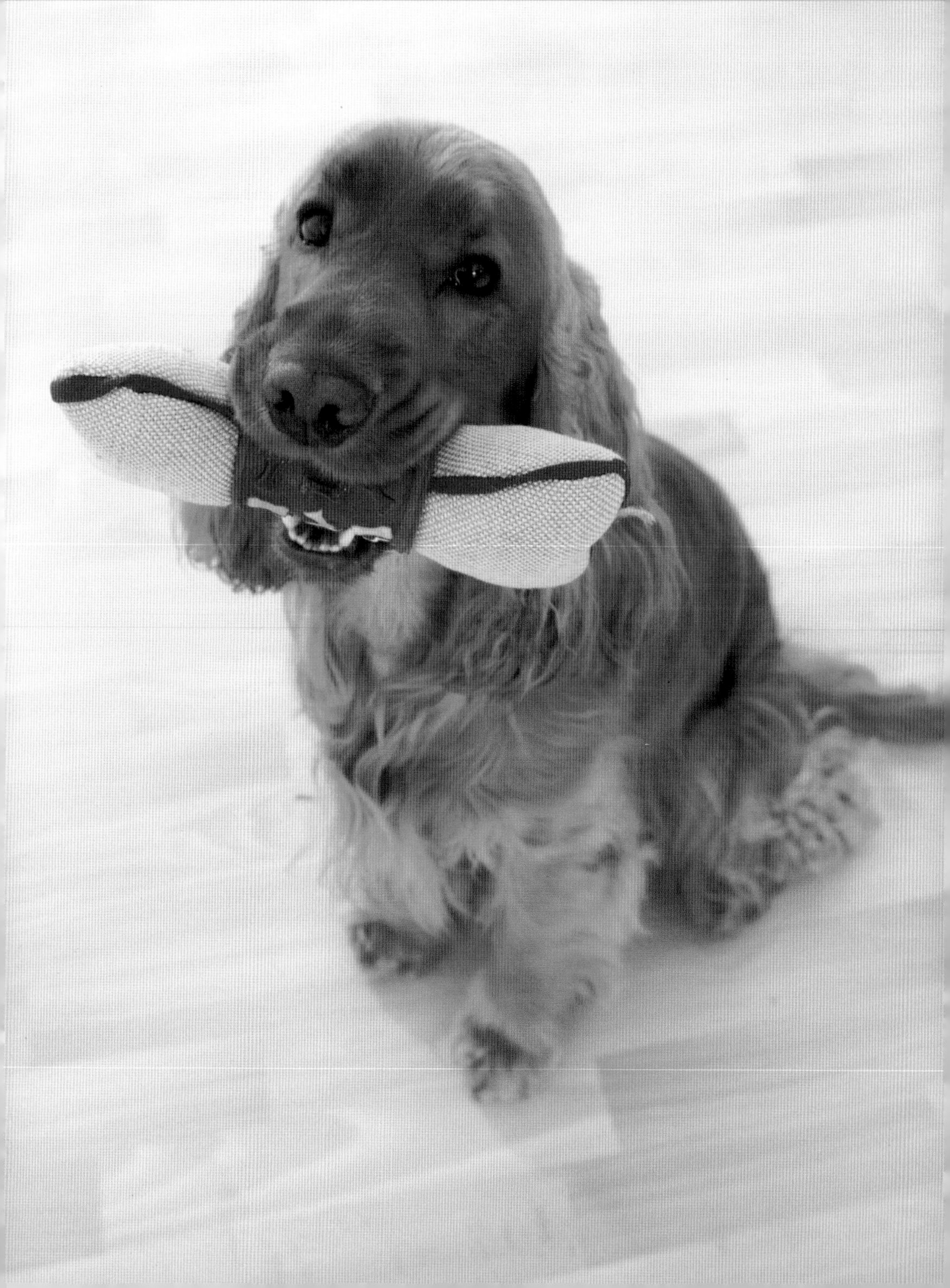

3

The daily mental workout

Boredom and a constant lack of stimulus will make your dog unhappy in the long run. A playful training program will keep him mentally and physically fit. As a bonus, you may discover that your dog has talents you've never suspected...

Learning exercises with zing

The following exercises will sharpen your dog's senses, attention and memory. They also make training easier, even though you might not think so at first.

Dogs that have learned to notice small signs are generally better at understanding what's wanted of them. Start off with simple tasks and gradually increase the degree of difficulty. After all, Rome wasn't built in a day. We too need some time before we learn to concentrate on what's essential.

The same goes for your dog. Doing the exercises should be fun for both of you, and they have nothing to do with obedience training. Instead, the goal here is to make your dog's playtime mentally challenging.

Observing accurately

Try to test your dog's powers of observation.

The cup test 1

◆ **Preparation:** Use two cups of very different shapes and sizes. Your dog should be sitting at a distance of three feet directly in front of you, observing you. Turn the cups upside down and put some dry dog food under one of them.

◆ **Performance:** Give the "search" command. The dog will go directly to the cup with the food, tip it over, and get his reward. Repeat this procedure several times. In all probability your dog has learned what the cup hiding the food looks like. You can check this. Switch the position of the cup and increase the distance. If your dog goes straight for the right cup, you knows he's grasped the concept.

The cup test 2

Now make the task more difficult. In this and the following exercise, don't allow the dog to see you hiding the food.

◆ **Preparation:** Line up five cups upside down in a row.

One of the cups is the same one from the first exercise.

◆ **Performance:** From the five cups the dog has to choose the right one. Once again, give the "search" command. It will take longer to find the right cup during this attempt, but it won't be a problem for most dogs. You can make it even harder by placing the cups at random around the room. What makes it harder? First, the dog has to understand that he has to find the right cup, and second, he needs more time to find it. For some dogs it isn't easy. You'll

need to practice this exercise a lot with your pet. Don't forget to praise him lavishly every time he makes the right decision.

The cup test 3

This exercise raises the degree of difficulty one step further.

◆ **Preparation:** Now put the food under a different cup. The cups are lined up in a row.

◆ **Performance:** Give your dog the "search" command. He'll go first, as he's learned, to

The cup test indicates how skilled your dog is in observing you.

TIP

Not in the mood for learning

Dogs, like humans, aren't always ready and eager to learn at any time on any day. Your dog's lack of interest is apparent when you're trying to show him what he has to do and he's not paying attention. His gaze wanders and he starts sniffing at the floor. When this happens, don't try to force him to be motivated. Instead, use gestures and a cheerful voice to encourage him to participate. If this doesn't work, don't interfere with his mood.

recently I was in a similar situation. I was staying in a hotel where the room door could be opened only by turning the key counterclockwise, in the "wrong" direction. How often did I fall into that trap within a week! This is just a small example. We have learned that our canine subjects differ from one another. Those that were curious and took the initiative in looking for the cup with the food learned more quickly than the rest.

Recognizing symbols

This is another exercise that tests your dog's powers of observation.

◆ **Preparation:** Take two identical food bowls and cover them with identical pieces of white cardboard. Draw a circle on one piece and a triangle on the other. Place the reward (food) in the bowl with the triangle.

◆ **Performance:** Give the "search" command. Since the difference between the bowls isn't as pronounced as in the cup test 1 (→ pg. 68), the dog will need more time to understand that the food is hidden under the lid with the triangle.

When the dog has learned that there isn't any food under the lid with the circle, change the task by drawing a square instead of a circle on the piece of cardboard and then covering the bowl with the square.

How does the dog react? It will hesitate to remove the lid with the square because it sees a new symbol. But it also knows that there isn't anything under the circle. So it chooses the square. Dogs that don't understand this and choose the circle will

the old cup, only to be disappointed that there isn't any food underneath. Now is where the dogs are divided into two camps. Some of them will search under the other cups. Others will stand there in a daze, overwhelmed by it all. If your dog belongs to the latter group, lift the cup hiding the food and show it to him. Now things get interesting. How long does your dog need to realize that the food can be found under another cup?

◆ **For your information:** We know from experience that dogs find this exercise difficult. Most of them will continue trying their luck with the old cup, to no avail. Unlearning something is not simple, as we know from our own experience. Just

Proper training

Training means variety, learning and physical exercise. But your dog isn't always in the right mood for it. You should observe the following rules.

■ Your dog should be well rested

You should practice training your dog only when he's fit. A dog that's tired or exhausted – for example after a long walk – will not feel like practicing. Take advantage of your pet's active phases.

■ Never train after meals

The dog should not start a training session on a full stomach. It's a well-known truism that "it's no use studying on a full stomach".

■ Tests are strenuous

The test phase shouldn't last more than 20 minutes. After a two hour walk our dogs were just as tired as after a 20 minute test.

■ Familiar surroundings

Carry out the tests in familiar surroundings. Avoid harsh light or loud background noises.

■ Familiar people

Make sure that only people your dog knows are present during the tests. Strangers will first have to get to know your pet.

■ Get your dog to ease up and relax

Before testing your dog, take him for a short walk. Always include breaks in your tests. During breaks, play with the dog and divert his attention. Don't repeat obedience exercises (→ pgs. 44-45) more than five to eight times

Here's a little exercise to relieve stress. Have the dog jump through your arms. (→ pg. 75).

1

2

The clue test

Is your dog able to understand the clues you give him? It's best to set up this exercise in front of a doorway. The dog is standing in front of you and can see everything you're doing. Place two cups upside down on two stools.

Hiding the reward

Now hide a reward under one of the cups, such as some dry dog food. The dog is watching what you're doing, but of course he's not allowed to get the treat. Now is when things get really interesting.

quickly get the message: Since there's nothing here, I'll choose the square to get the treat.

◆ **Increasing the difficulty:** You can now challenge your pet's attention by making the exercise harder, little by little. Change the square into a hexagon, an octagon and then a polygon. You already notice that a polygon looks very much like a circle. The shapes are very similar, and the dog finally is unable to notice a difference. His reaction is surprising. He'll look to you for help. If you don't give it, he'll then lie down between the food bowls and begin to

scratch or lick himself. This is typical behavior in a dog that's been overtaxed. He gives up after several attempts. I wouldn't let things get so far, since by this time the dog isn't getting any enjoyment from the exercise.

The clue test

This test shows how well your dog is able to understand your clues (see photos above). Is your dog a little genius?

◆ **Preparation:** Place two stools next to each other, with an inverted cup on each

3

Blocking his view

Put up a sheet or towel between you and your dog (you can fasten it to the doorway with thumbtacks). Hang it high enough to hide the two stools from your dog's view. He'll probably look a little puzzled.

4

Leaving a clue

Now switch the cups and put a tennis ball or other object on top of the cup that has the food. Remove the sheet. Will your dog understand the clue and thus go straight to his reward?

one. The dog is sitting about three feet in front of them. You then hide a reward under one of the cups. On top of the cup you then place an object (a tennis ball, for instance). At the beginning the dog should see you placing the object on top of the cup with the food. The two stools are then hidden behind a sheet, the cups are switched, and the tennis ball is placed on the cup that contains the food.

◆ **Performance:** Now remove the sheet and have the dog go for his reward. If he goes for the right cup, tilt the cup over and give your pet his treat. If he's made the wrong choice, he doesn't get anything. After two or three attempts your dog has finally realized that the food is under the cup that has the object on top.

◆ **For your information:** A few of the dogs we tested made up to ten attempts before they understood the clue. It really isn't all that easy, since a possible clue could be either the owner putting an object on a cup, or the object itself. One of our six canine subjects failed the test completely, so don't be disappointed if your dog has difficulty with it. Maybe he's better at other tests and can master them brilliantly.

Exercises to keep your dog fit

Quick in its thinking, but also quick on its legs – that's the key to your pet's well-being. It's the combination that counts.

Along with mental stimulus, your dog needs an adequate amount of physical exercise. Exercise strengthens the heart, circulation, musculature and brain. Why the brain? Because it controls complex motor functions. Think of a tennis player, and how quickly he has to react when hitting the ball back and forth. His eyes can hardly follow the action. In a fraction of a second he has to decide how to hit the ball. This is when the brain is performing at top capacity. It's no wonder that constant training is required. Obviously, your dog isn't going to become a tennis player, but he should also learn physical agility, whether he's a dachshund or a Saint Bernard. That's why I recommend an agility training run. For a lot of dogs, going through this run is great fun.

For your information: The performance of the agility exercises depends, of course, on the size and breed of the dog. As a rule, smaller dogs perform better. You'll notice I was thinking about Wisla, my Saint Bernard. Plenty of exercise has turned her into a sure-footed big dog. And as a special bonus, she's much more confident when approaching new things.

■ Balancing

Exercise: Your dog learns how to keep its balance on a log.

Performance: Hop onto a log that's lying flat on the ground and use a coaxing voice to encourage your dog to join you. If your voice alone doesn't do the trick, try tempting him with a treat. Don't use a commanding tone of voice, since you want the mood to be playful. When the dog jumps onto the log, praise him generously. Repeat jumping onto the log several times, so that the dog loses his insecurity. When this is successful you can start balancing on the log.

■ Jumping

Exercise: Your dog learns how to jump over logs or streams

Performance: Start jumping for fun over a small log or narrow stream. Encourage your pet to copy you by using a coaxing voice or a treat. When the dog follows you, don't forget to praise him lavishly.

Increasing the degree of difficulty: Do this by gradually increasing the width of the obstacle he has to jump over.

■ Crawling through narrow spaces

Exercise: Get your dog to crawl through a narrow tunnel. This can be, for example, a play tunnel made out of nylon or an existing obstacle, such as a small drainage tunnel.

Performance: Place a treat in the middle of the tunnel. Use a tempting voice to get him to fetch the treat. If the dog is afraid, try again using a bigger tunnel. If he can crawl through that, go back to the original tunnel.

■ Climbing stairs

Exercise: Your dog learns to walk up steep, narrow stairs.

A dog will learn to jump over a log or a small stream in no time.

Performance: First put your dog on the leash and then, together with him, climb up a flight of stairs. Pet him and use a soothing, gentle voice to say, "good dog, good boy", so that he knows he has your trust.

■ Jumping through your arms

Exercise: Have your small or medium-sized dog jump through your arms. To do this, you'll have to kneel down and form your arms into a circle next to your body. First make the circle close to the floor.

Performance: Have a friend use a treat in order to coax the dog through your arms. As soon as the dog has understood what's wanted of him, slowly raise the circle from the floor. A treat is used once more, this time to get him to jump through the circle.

Balancing on a log helps train your pet's sense of equilibrium.

1

2

Recognizing signals

How closely does your dog observe you? Do the test. Your dog is sitting and facing you. In one hand you're holding a treat, which he can see, and you're looking straight at him. He's not allowed to go for the treat.

Finding the treat

Bring your hands behind your back and switch the treat to the other hand. Close this hand to hide the treat. Now stretch your arms out to the sides, turn your head towards the hand with the treat, and give the "search" command. What happens now?

The smelling test

You haven't forgotten that dogs are champions when it comes to detecting the right smell. It was a pleasure for me to find out how well my dog recognized my scent.

◆ **Preparation:** Take two identical feeding bowls and cover them with identical pieces of white cardboard. Now put a scent marker on the cardboard. You can rub salami or perfume on one piece, and on the other you leave your own scent. To do this, hold the piece of cardboard five to ten minutes in your hand and then rub your hand several times on it. Put the reward (food) in the bowl that has your scent.

◆ **Performance:** Now have your dog sniff your hand thoroughly. Then give the "search" command. The dog will go straight to the food bowls, immediately detect your scent, and take his well-earned reward from the bowl.

◆ **Increasing the difficulty:** Use several food bowls and give them all different scents. Then let your dog search for the scent you've chosen.

Recognizing symbols

Once again, we test your dog's attentiveness. How closely does he observe you?

◆ **Preparation:** The dog is sitting about three feet away and facing you. You're holding in one hand a treat that he can see.

◆ **Performance:** Now the actual test begins. Behind your back, hold the treat in one hand, keeping the hand closed so the dog can't see the treat. Then stretch your arms out to the sides, turn your head towards the treat and give the "search" command. Dogs that have always keenly observed their owners will grasp the meaning of the turned head after four or five times. Other dogs will need considerably more practice. Of course, a dog's ability to comprehend depends on its natural talents.

◆ **Increasing the difficulty:** You can now refine the test by not turning your head anymore, and instead looking directly in the dog's eyes while keeping your head still. Only the movement of your eyes indicates the hand that holds the treat. Your dog will learn your eye language relatively quickly. We were pleasantly surprised. The test shows that the dog is observing you precisely. He's trying to read our eyes to find out what we want.

The pointing test

Can your dog understand you without you speaking? If that's the case, the two of you have an especially tight bond.

◆ **Preparation:** The dog is sitting facing you about six feet away. To your left and to your right, at a distance of about 15 feet, there are two feeding bowls. One of the bowls contains a treat and the other is empty, but the dog can't see which bowl has the treat.

◆ **Performance:** Now stretch your arm and index finger out in the direction of the bowl with the food. Without hesitation, your dog will follow your indication and head eagerly for the treat.

TIP

A reward makes the difference

Your dog should not be fed for at least two to three hours before doing any of the learning and reasoning exercises. Otherwise, a food reward won't be attractive enough. On the other hand, if your dog is too hungry, he won't be able to learn and think logically. After he's successfully finished the tests reward him with praise, petting and a treat. If the dog fails, calmly start over without saying anything.

◆ **For your information:** Dogs make very few mistakes. They immediately understand the gesture. This is in contrast to wolves, their ancestors, who don't notice the hint and will make a mistake 50 percent of the time. Scientists at Budapest University have gone a step further using a similar, but much more difficult, experiment: The owner holds a stick behind his back and points it towards the food bowl. It's amazing how some dogs understood even this gesture.

Learning Ricostyle

Can you still recall Rico (→ pg. 55)? Then maybe you can find out if your dog, too, is a little "Einstein".

◆ **Exercise:** Have your dog bring you a stuffed toy by name. More precisely: "Teddy, bring me the turtle".

◆ **Performance:** Practice this fetching exercise repeatedly over several days as part of playtime. Wait until you're certain that your dog has mastered the task before adding another stuffed animal, let's say in this case, a lion. The first five to ten times continue calling for the turtle, and then call for the lion. The dog will be confused and won't know exactly what to do. That doesn't matter. Just pick up the lion and put it in the dog's mouth, saying repeatedly "lion". After a few days, the dog will associate the word with the object. That's when you put him to the test: With the lion and the turtle lying in front of the dog, you order him to bring you the turtle. If he makes the right choice, he's learned the lesson.

Give a toy a name and have your dog bring it to you.

◆ **Increasing the difficulty:** If your dog can match your words to two stuffed animals, you can expand his "vocabulary" to other stuffed toys.

The shell game

The aim of this game is once again to test how well your dog can pay attention.

◆ **Preparation:** Take two identical cups or food bowls, turn them upside down, and place a treat under one of them. The dog is watching as you do this.

◆ **Performance:** Now, with your hands push the cups around, making overlapping circles. The dog's job is to follow with his eyes the cup that's hiding the treat. You should suddenly stop moving the cups when they're lined up horizontally in front of you. Now give the "search" command. Will your dog find the cup with the treat?

◆ **For your information:** During testing, our dogs were curious and focussed their attention on following the movements of the arms and hands. The success rate was 50%, which is to say that it was probably luck when the dogs made the correct choice. The dogs were just as disappointed as we were. What was the reason for the sobering results? We tried slowing down our circling movements. Again, there was curiosity and wonder on the part of the dogs, but no improvement in the results,

TIP

Young dogs

Young dogs (under six months old) are only too happy to play, and they're not yet able to concentrate on given tasks – apparently just like little children. You have to take this into account. Let your puppy play with other dogs, and stimulate its senses by taking it on walks where it will learn to deal with its surroundings through playing. Of course, each dog develops differently.

even when we turned over the cups so that the dogs could see and smell the food. Each dog was visibly confused. What's your dog's success rate?

The cheating game

Do you still remember Gina, the dog that went for the forbidden treat while her mistress's face was hidden from her by a newspaper she was pretending to read (→ pg. 60)? At the Max Planck Institute I got to witness a similar test, one that you can easily duplicate. Maybe your dog is also a small-time cheater.

◆ **Preparation:** Sit down in a chair and make yourself comfortable. Your pet is at a short distance from you and has a good view of you. Lying on the floor in front of both of you, about six to ten feet away, there's a tasty treat. Give your dog the "down" command and forbid him to grab the treat. Any attempt to get near it is met by a sharp "stop!".

◆ **Performance:** Close your eyes. What happens now? You can get a description of your dog's behavior from somebody who's watching from a distance. Probably the dog will start looking back and forth between you and the treat, giving the impression that he's testing his owner's reaction. No reaction from you is taken as a green light for him to go for the treat. But before this happens the dog is restless, often scratching himself, and the sight of the treat makes his mouth water. He'll walk slowly towards the temptation without losing sight of you. One can see that he knows that he's cheating.

Tricky tests of logic

My team and I carried out the following tests in order to get an idea of canine intelligence. You can try them out on your own dog. Maybe you've got a four-legged genius on your hands.

Fishing for food 1

◆ **Preparation:** Instead of a cage we used a bicycle basket, flipped over and nailed to a wooden board. We placed a treat in the basket, on a narrow strip of cloth that had one end sticking out of the basket. It was the dog's job to pull on the strip of cloth in order to get to the treat. It's a difficult task, since the dog has to realize that the food is on the cloth, and that he then has to move the cloth (see photo opposite).

◆ **Performance:** A lot of the dogs tried their luck using raw force. They pawed and scraped frantically at the basket and finally gave up, exhausted. The smarter dogs also started off by scraping, but then they circled the basket and deliberately pulled at the cloth. I could clearly see that they were thinking while doing this.

Fishing for food 2

This test is suitable only for dogs that have already mastered the previous test.

◆ **Preparation:** Setting up is similar to the previous test. It's made more difficult by adding a second strip of cloth that protrudes out of the opposite end of the basket. Only one strip, however, has food on it. The dog has to look carefully in order to make the right decision.

◆ **Performance:** Dogs find this test difficult. They simply start pulling at either of the cloths. It wasn't uncommon for a dog to fail at 20 attempts. The smart dogs – or maybe they were just good observers – were able to pull the right cloth. It was an enormous achievement. They circled the basket before decisively pulling the right cloth.

Fishing for food 3

◆ **Preparation:** We wanted to really challenge the observational and reasoning skills of our candidates. Setting up the experiment was the same as before (→ fishing for food 1, pg. 80), but now there were two treats in the cage. One of them was on a strip of cloth, and the other was next to an identical looking strip. The treats were covered with small transparent bowls. Without the bowls the dogs were helpless. Apparently, the smell of the treats was so spread out that it prevented the dogs from making the right choice. The bowls restricted the smell.

◆ **Performance:** The test subjects were indeed able to get the food. The dogs' performance was all the more impressive when you consider what was going on in their minds. They had to understand that only the food on the cloth could be moved

A bicycle basket, strips of cloth and treats are turned into a test of logic.

1

2

Fishing for food 4

With some bricks and boards you can amaze your dog. While your dog's looking, drop a treat into the crack between the boards. There's no question that your pet will want to get this reward.

Good advice is hard to come by

But how to get to the tasty treat? The dog's muzzle doesn't fit through the narrow gap. Now what? Most dogs are helpless at first. But the "brains" among them soon find out how to "win" the treat.

– that is, they had to make the causal connection between motion, cloth and food. Just how complex this problem is can be seen in babies. They're not able to solve such problems until they're a year and a half old. Perhaps dogs also have an idea of everyday physics.

Fishing for food 4

◆ **Preparation:** You'll need some bricks and boards. Lay down two bricks lengthwise end to end, with another two bricks on top of them. Then set up an identical parallel row of bricks eight to twelve inches away. Lay two boards on top of the bricks to create a "house". Leave a narrow gap between the boards through which you can drop some food. If you want, you can secure the boards by putting bricks on top of them.

◆ **Performance:** With your dog watching you, drop a treat through the gap. The dog wants to put his head through the gap, but that doesn't work. He has to find another way to get to the food. The solution is achieved by reaching with his paw from the side, instead of from above, to "fish" out the food.

3

4

hat's the way

ince the treat fell through the gap,
ou'll have to look for it on the floor. It's
o problem at all to get to the snack by
oing through the open side of the con-
truction.

It gets harder

Now things are a bit more difficult. The
"house" has gotten taller and the treat
has fallen onto the "second floor". After
a bit of thinking the dog has reached a
solution. Don't look on the floor, look
one level higher.

In our experience, about 50 percent of the
dogs solved this not-so-simple task after a
few tries.

◆ **Increasing the difficulty:** It gets harder
for the dog when you put up a two-story
"house". As before, drop the food through
the gap. This time it lands "upstairs"
instead of on the floor.

The dog will first search on the floor, find
to its surprise that there's no treat there,
and then take another look through the
gap. After a short while it grasps the
problem and fishes the food from the
second level.

The secret of the lid

◆ **Preparation:** Take a dog bowl and cover
it with a cardboard lid. Your dog is sitting
three feet away from you and can see what
you're doing. Now lift the lid, put some
food in the bowl, and put the lid back on.

◆ **Performance:** Now give the "search"
command to have the dog look for the
food. Most dogs will rush to the bowl, push
the lid away with their muzzles, and eat the
food. It's no big problem for the dog, then.
He's realized that to get to the food he has
to push the lid off.

◆ **For your information** Cats fail this test. We tried it on over 60 cats, always with the same results. A requirement of this test is that the animal has never before confronted this, or a similar, problem. Among cats there may be high achievers, but the majority fail this test. All sorts of theories went through our minds. Could this possibly be linked to evolution?

We tested wolves at the University of Kiel under Dr. Dorit Feddersen-Petersen. The wolves approached the bucket with the lid, sniffed everything, circled the bucket and urinated on the lid. They thus claimed the object for themselves, but didn't solve the problem. Dingoes, the once-domesticated wild dogs of Australia, did remove the lid, but they needed a lot more time than a normal house dog. As far as cats go, we still don't have an answer, but it seems that they view the bowl and lid as one object, and can't grasp that they are actually two things. I've made this brief excursion into the animal kingdom simply to demonstrate how difficult it is to make conclusions about animal intelligence.

Spatial perception

◆ **Preparation:** Take two sheets of clear, hard plastic, such as plexiglass, and prop them up end to end to create a V-shape. Leave a small opening at the bottom of the tip of the V through which you can pull a small piece of meat attached to a string.

◆ **Performance:** Have the dog sit between the two plexiglass sheets, facing the opening. The piece of meat with string attached is lying only an inch or two from his muzzle. Now, working from the outside, pull the meat by the string out of the enclosure.

The dog will follow the string up to the opening, but he can't go any further than that. He'll bark and whine and try to get at the meat with his muzzle or paws, but that's the wrong strategy. To get to the meat he only has to walk around the plastic enclosure – something not all dogs can manage. In any case, once you've shown your dog the solution to the problem, he won't forget it so quickly.

◆ **For your information:** Just what's going on in the mind of a dog that's not able to solve this apparently simple task? Behavioral scientists at the University of Budapest carried out similar tests with surprising results. Instead of plastic, they used

To test your dog's spatial perception you'll need two sheets of plexiglass, a string and some food.

1

Spatial perception

The sheets of plexiglass (or wire netting in frames) are propped up in a V-shape so that there's a small opening at the tip. Now put a tasty snack, attached to a string, in the enclosure. Your dog is sitting inside it.

2

Not at all easy

When the dog then tries to get to the treat, pull the string out through the opening. The reward can't be reached, since only the dog's muzzle can get through the narrow gap. The dog has to find another way. And, as you can see, he's found it.

3

Thinking it through pays off

Naturally, now comes the hard-earned reward. Finding the right way wasn't easy at all. Your pet will first try to "soften you up" by whining and barking, but then he'll "turn on" his brain power and finally walk around the enclosure.

a V-shaped fence without an opening and placed the food outside the tip of the fence. The dog could thus easily smell and see the food.

The dogs had no problem with the task, and they walked around the fence. But when the positions were switched, with the food placed within the fence and the dog standing outside, the dogs then had great difficulty in reaching the food, even when the experiment was repeated several times. They became much better at it after they saw a human do it, whether this was their owner or a stranger.

Finding the shortest way

◆ **Preparation:** You'll need an approximately 30-foot fence that has an opening at either end, and a stick for the dog to fetch.
◆ **Performance:** Stand in front of the fence with your dog. Throw the stick to the other side of the fence. Usually a dog will run to the fence, and start barking and digging. It doesn't realize that it has to run around the fence. Only the smartest dogs will immediately run around it.

Lead your dog along the fence and show him how to get to the stick. With the next throw he'll know what to do. The next time, place yourself closer to one of the openings when you throw the stick. A smart dog will then choose the nearest opening to get to the stick, while a slower dog will take the same old route.

Test how quickly your pet finds the shortest way around the fence.

The top hat test

◆ **Preparation:** Put together a "top hat" out of cardboard, making the brim especially wide. How big the hat is depends, of course, on the size of your dog. With your dog watching, put a treat under the hat.
◆ **Performance:** Now command the dog to get the treat. He'll try to knock the hat over with his paws or muzzle, but unsuccessfully, since he's standing on the brim of the hat. It takes a long time before the

dog tips the hat over, either by chance or deliberation.

◆ **For your information:** The twenty dogs we tested needed 20 minutes, on average. We think that some of them figured out the problem, while the others worked by trial and error.

The half top hat

◆ **Preparation:** This is the same test as described above, except that instead of the brim going completely around the hat (360 degrees), it goes only halfway around (180 degrees).

◆ **Performance:** Our belief that some of the dogs did indeed grasp the problem was

reinforced when we saw some of our subjects approach the hat from the side without the brim. In order to rule out mere coincidence, we rotated the hat. The dogs immediately chose the side without the brim.

What is a large amount?

This experiment tests whether your dog has an idea of quantities.

◆ **Preparation:** Drop differing amounts of treats or food morsels into two food bowls. Then place a piece of cardboard on each bowl. Your dog is sitting about three to six feet away and watching you.

◆ **Performance:** Command your dog to go for his reward. Will he go for the bowl with the most food? The second time around, switch the bowls.

◆ **Tip:** The experiment may seem simple, but it has a few catches. Dogs often prefer one side to another. They decide in advance where they're going, without paying attention to quantities. A possible explanation is that the quantity isn't important at the moment, since they're going to get both rewards anyhow. Either choice is successful and they don't need to think about it. And just how can we challenge the dog's attention? By sometimes not dropping food into one of the bowls, willy-nilly. This will make the dog watch you more closely in order to determine which bowl the food is going into. Interestingly enough, our test dogs were able to differentiate among amounts between one and five.

INFO

Thinking takes energy

Each series of tests should not take more than an average of 20 minutes. The dog is then exhausted. You should allow the dog at least a one hour break between the series of tests. More than two test series is too many. If the dog refuses to work, you know it's time to stop. Let him then choose if he'd like to go for a walk or take a short nap instead.

Memory training

Dogs have a good memory and can be trained just like in humans. You can test how well your dog remembers things.

■ In the labyrinth of the past

Getting by in a world without memory is impossible. Imagine you had to live without memories. The idea itself is awful and frightening. Animals also remember. Just like us, they have a short-term and a long-term memory. Some animals are even masters of memory.

Learning and reasoning are impossible without memory. For the duration of our lives, experience and memory function together and are inextricably linked to each other. Each new flash of thought is presumably bound to memory.

I don't know whether dogs actually have a proverbial "elephant's memory", and I also don't know of any scientific research on the subject of canine memory, but from personal observation I can say that dogs have a good memory. They bury their bones and dig them up weeks later.

The dogs that had participated in our experiments remembered procedural details twelve months later. They began the same experiment as precisely as if they had done it yesterday. The results were also almost exactly identical to those of the year before. The dogs hadn't forgotten anything.

■ A good memory

Just how good a dog's memory is can be easily observed in everyday situations. After going on the same walk four or five times your dog knows if you're going to turn to the right or left. You can test this simply be letting your dog walk a few feet ahead of you. You'll notice how he trots along the familiar route. Of course, this isn't absolute proof. He could just be following the scent of yesterday's walk. Ruling out that possibility would require a good deal of work, but another clue is given by the dog's behavior: If he's walking casually without sniffing the ground, this indicates his memory is at work.

I don't know if all dogs have good memory, but I imagine that there are the same variations as in humans. Anything else doesn't make sense to me, since learning and reasoning depend on personality. And since reasoning and memory work together, I assume that each dog's memory is individual.

■ Memory exercises

Can memory be trained? This is possible to a certain extent in humans, and in dogs I believe this is also the case. Otherwise, it would be hard to explain the achievements of Rico, the celebrated border collie (→ pg. 55.) Why don't you try testing this out on your own dog?

Exercise 1: A dog that's good at fetching can be taught to associate a word with an object. This means it has to memorize which object belongs to which word.

Here's how it goes: Show your dog the leash, and say "leash". Then put the leash in his mouth and repeat the word "leash".

Teach your dog to associate an object with a word.

Do this every day for several days. Now take a different object and give it a different name. Follow the same procedure with the new word. Now comes the real test: Has your dog forgotten the word "leash" when you say it? Each of my dogs remembered it immediately, and they all eagerly brought me the leash

Exercise 2: During a walk, with your dog watching, hide several objects you've brought along and give them names. Make sure that the objects are attractive to the dog, or else he won't notice them. But don't let him take them.

The next day, take the same route and have your dog search for the objects. He won't hesitate to start looking. This type of memory training is great fun for dogs.

Searching for and recovering objects. It's a great way to train your dog's memory.

Index

Page numbers in **boldface** indicate illustrations.

The author

Dr. Immanuel Birmelin has devoted over 25 years to investigating the behavior of animals in zoos, circuses and the home. He has intensively studied pet behavior and was able to demonstrate through tests that dogs have the ability to reason. We must therefore realize that dogs are not weak-willed servants of man, but rather thinking, sensitive fellow creatures. The dog owner who accepts this fact will find it considerably easier to teach his dog many things. The need for intensive drilling is not as great as previously assumed. This was the subject of Immanuel Birmelin and Volker Arzt's film, "Dog or cat – which is smarter?" that created a sensation in 2003. This was followed by guest appearances on television talk shows. In addition, the author is active as a scientific advisor for animal film productions and as an authority on the correct care of animals.

Other great Hubble & Hattie titles –

978-1-845842-91-8

978-1-845842-92-5

978-1-845842-93-2

978-1-845842-74-1

– with many more to come!

Back cover: 'Sunbeam Hattie'

Photo credits
Aschermann: pg 36; Corbis/Williams: pg 17; Corbia/Zefaimages/Le Fortune: pg 9; Getty Images/Fujiwara: pg 10; Getty Images/Neo Vision: pgs 4, 23; Giel: cover, pgs 12, 22, 30, 35, 39 left, right, 47, 48, 49 left, right, 51 bottom, 53, 57, 62 left,, right, 73 left, right, 75 top, bottom, 76 left, right, 81, 82 left, right, 83 left, right, 84, 85 top, centre, bottom, 89 top, bottom, spine; Juniors/Wegler: pgs 8, 59 top; Schanz: pg 42; Steimer: pgs 54, 59 bottom, 80, 86; Wegler: pgs 2 top, bottom, 6-7, 14, 14 left, right, 19 top, bottom, 20, 21, 24, 26, 28, 32-33, 40, 41, 45 top, bottom, 51 top, 52, 60, 66-67, 68, 71, 78

About the author

Dr Immanuel Birmelin has devoted more than 25 years to investigating the behaviour of animals in zoos, circuses, and the home. He has extensively studied pet behaviour, and was able to demonstrate through tests that dogs have the ability to reason, and are therefore not weak-willed servants of man, but thinking, sensitive fellow creatures.

The dog owner who embraces this fact will find it considerably easier to teach his dog many things; the need for intensive drilling is not as great as previously assumed. This was the subject of Immanuel Birmelin's film *Dog or cat – which is smarter?* which created a sensation in 2003, and was followed up by guest appearances on television talk shows.

The author is also active as a scientific advisor for animal film productions, and as an authority on the correct care of animals.